Building with Lime-stabilized Soil

Praise for this book

'This book extends our understanding of conservation; linking traditional building materials and methods from the past, with the creation of modern sustainable homes in response to the acute impacts of climate change. It offers inspiration and empowerment to both the vernacular building owner and international aid organizations.'
Nichola Tasker, Chair of the Society for the Protection of Ancient Buildings

'Architects take note, read and learn, and take these techniques out into your practice. Not just in the developing world – developed economies need this just as much if not more. It is the next big thing in reducing your carbon footprint (or more accurately it was something vernacular builders knew well, but we are only now recognizing its importance). What could be more compelling than using the earth and rocks dug out of the construction site to make into architecture – resulting in zero transport miles?'
Charles Parrack, Course Leader Professional Masters in Architectural Design, Oxford Brookes University

'Sustainable small-scale and easily maintained reconstruction methods that can help rebuild disaster-affected communities are very welcome in the humanitarian context, all the more so within the context of the Sustainable Development Goals and the drive to greener, more climate-friendly solutions. This useful primer offers tried and tested techniques that can empower communities and increase their resilience and recovery capacities.'
Marianne Farrar-Hockley, European Commission, European Civil Protection and Humanitarian Aid Operations

'Finally a user-friendly guide to lime and earth in building, well illustrated for training purposes and detailed enough to cope with a wide range of situations. Bee and Stafford have been working with these materials for a long time and understand the needs of builders and designers as well as the practical issues which face anyone trying to build with local materials, to do local field testing and give practical solutions which can be widely applied. This is a book for trainers, practitioners, organizations, regulators and funders to understand the issues in choosing local building materials and skills, and moving away from environmentally damaging and expensive industrialized solutions.'
Rowland Keable, CEO, Earth Building UK and Ireland

'From home improvement to a mass resilient housing programme – this book is essential reading for anyone working on any project where using lime-stabilized soil is an option for any part of the construction. Whether experienced in the use of lime, or merely curious, there is clear and well-explained technical content for all. Built from tried and tested practice, and founded in solid research, this guide is not only for those with a passing interest, but also for practitioners seeking clear instruction on how to do everything from testing soils to finishing renders. In brief, it is the "go to" reference on the subject. I hope, and expect, that readers will feel both inspired by what is possible and guided by this book to build more sustainable and cost-effective structures across the world.'
Joseph Ashmore, Head of Shelter and Settlements, Global Shelter Cluster, International Organization for Migration

'I first encountered Stafford Holmes when I was working on alternative cements in ITDG in the late 1970s, and I am delighted to see that the approach we advocated at the time has been further developed both for historic buildings and for recent post flood and earthquake projects in Pakistan and Nepal. This book, based on more than 40 years of experience with the material, is bound to become the standard document on how to select, test and use lime-stabilized soil. It is comprehensive and well written and it has been appropriately illustrated.'
Robin Spence, Emeritus Professor of Architectural Engineering, Cambridge University; Director, Cambridge Architectural Research Ltd

'For over 20 years Bee Rowan and I have been colleagues through our work with natural building in Europe and the US. In 2018 Bee diligently began educating myself and local builders in how to use lime-stabilized soil for durable renders, plasters and floors for the first straw bale house in Nepal. This included quality-testing burnt lime (quicklime), slaking quicklime for lime putty, determining the optimal lime proportion for the local soil, and finishing the walls with a stunning lime wash! This book can do the same and more for its readers. In *Building with Lime-stabilized Soil*, Stafford Holmes and Bee Rowan fill a void of historical and practical information in the use of lime-stabilized soil in buildings. They revive this proven way of building with both centuries-old empirical knowledge and current scientific understanding. Never has the global need been greater to (re)activate the environmental and economic benefits of building with lime-stabilized soil.'
Martin Hammer, Architect, Co-director of Builders Without Borders

'*Building with Lime-stabilized Soil* is the book I have been looking for, and which we now need to consider translating into Nepali and other languages. The subject is presented in a very helpful way and allows for low-income countries like Nepal – historically familiar with lime, the knowledge of which is mostly lost and subsumed within cement concrete construction – to rediscover how to make the best use of lime, cost-effectively and with low environmental impact. Nepal has an abundance of lime, currently devoured by an ever-expanding cement industry, whilst the corresponding mass extraction of sand is wholly unsustainable – destroying our rivers and their rich biodiversity. This book is so complete that it introduces lime to its readers in a clear manner, whilst offering amply illustrated technical and procedural knowledge on how sustainable and cost-effective construction can be achieved through stabilizing soils with very small amounts of lime, thereby reviving the use of soil – found in abundance, and cheaply – in developing countries. Backed with wider initiatives of practical training programmes, this book will prove to be a milestone in reference material for transforming the current global unsustainable building and infrastructure development approaches, to those greener, more cost-effective and culturally rich.'
Shuva Sharma, CEO of Scott Wilson Nepal, engineering dedicated to green infrastructure development

'I can only strongly recommend this important addition to our understanding of the success over millennia, and the future potential, as a sustainable and long-lasting building material of lime-stabilized soil, not only in the context of climate warming, but in that of human health and happiness. The ability of relatively small volumes of pure lime to beneficially transform the workability and performance of an earth, or loam, mortar had long been understood by the crafts, but has been too long forgotten or ignored by the construction and conservation sectors alike. This book – and the inspiring achievements of local people and communities which form the context of its creation – deserves to be widely read and, as importantly, to be acted upon.'

Nigel Copsey, conservator, stonemason and author

'Building with lime has been used and forgotten in almost every continent around the world. After hundreds of years, many structures are still standing, from Iraq to the UK, notably in surviving Roman architecture and infrastructure. Whilst there is less historical evidence for lime stabilization of soil, this book gives compelling examples from an equally wide-ranging geographical spread. Lime stabilization of soil is a cheaper and more environmentally friendly alternative to many other construction methods, particularly the use of Portland cement, and can be far easier to integrate or reintegrate sustainably into building cultures. The unique and hydraulic properties of lime-stabilized soil often make it the most appropriate option, including for the strengthening and protection of buildings from slow onset or prolonged flood. Stafford Holmes and Bee Rowan offer a well-illustrated practical book, drawing upon extensive experience and sources. Their efforts build upon work by other natural building experts, including John Norton, Hugo Huben and Hubert Guillaud, celebrating also continuity with Practical Action over this environmentally sensitive approach to construction, an option for consideration in almost every building project.'

Tom Corsellis, Executive Director, Shelter Centre, Geneva

'Stafford Holmes and Bee Rowan are highly respected in their fields who each have taught at the Centre for Alternative Technology for many years and now share their extensive knowledge, skills and experience in this detailed, wonderfully illustrated and accessible new book. Our global climate challenged society urgently needs regenerative solutions that both mitigate impacts and protect the most vulnerable. This book provides a positive and applied contribution to literature on the many benefits of building with low-carbon, natural materials, including the environmental benefits through their life cycle and improved building performance. The work clearly communicates the importance of learning from and working with vernacular building knowledge, enhancing earth-building practices through findings from the authors' scientific research to create durable solutions for historic buildings, rural communities and within the shelter sector – highlighting important interactions between global and local perspectives to create sustainable solutions.'

Tim Coleridge RIBA, Programme Leader, Sustainability and Adaptation, Graduate School for the Environment, Centre for Alternative Technology

Building with Lime-stabilized Soil

Stafford Holmes and Bee Rowan

Practical Action Publishing Ltd
27a Albert Street, Rugby
Warwickshire, CV21 2SG, UK
www.practicalactionpublishing.com

© Stafford Holmes and Bee Rowan, 2021

The moral right of the authors to be identified as editors of the work and the contributors to be identified as contributors of this work have been asserted under sections 77 and 78 of the Copyright Designs and Patents Act 1988.

All rights reserved. No part of this publication may be reprinted or reproduced or utilized in any form or by any electronic, mechanical, or other means, now known or hereafter invented, including photocopying and recording, or in any information storage or retrieval system, without the written permission of the publishers.

Product or corporate names may be trademarks or registered trademarks, and are used only for identification and explanation without intent to infringe.

A catalogue record for this book is available from the British Library.

A catalogue record for this book has been requested from the Library of Congress.

ISBN 978-1-78853-001-9 Paperback
ISBN 978-1-78853-000-2 Hardback
ISBN 978-1-78044-700-1 eBook

Citation: Holmes, S. and Rowan, B. (2021) *Building with Lime-stabilized Soil*, Rugby, UK: Practical Action Publishing <http://dx.doi.org/10.3362/9781780447001>

Since 1974, Practical Action Publishing has published and disseminated books and information in support of international development work throughout the world. Practical Action Publishing is a trading name of Practical Action Publishing Ltd (Company Reg. No. 1159018), the wholly owned publishing company of Practical Action. Practical Action Publishing trades only in support of its parent charity objectives and any profits are covenanted back to Practical Action (Charity Reg. No. 247257, Group VAT Registration No. 880 9924 76).

The views and opinions in this publication are those of the author and do not represent those of Practical Action Publishing Ltd or its parent charity Practical Action. Reasonable efforts have been made to publish reliable data and information, but the authors and publisher cannot assume responsibility for the validity of all materials or for the consequences of their use.

Cover photos by the authors
Illustrations: Juliet Breese, www.deaftdesign.co.uk
Typeset by vPrompt eServices, India

Contents

Cover photos	x
About the authors	xi
Foreword *Magnus Wolfe Murray*	xiii
Preface	xvi
Acknowledgements	xvii
Abbreviations	xviii

1. **Lime-stabilized soil: an invaluable resource** — 1
   - 1.1 Introduction: economic and environmental benefits — 1
   - 1.2 History and durability — 3
   - 1.3 Ecological advantages — 5
   - 1.4 Health and sustainability — 6
   - 1.5 Low-cost construction — 8
   - 1.6 Diversity of building elements — 9
   - 1.7 Disaster recovery and rural development — 9
   - 1.8 Simple field-testing methods — 9
   - 1.9 Established and current use — 10
   - 1.10 National standards for lime-stabilized soil — 11
   - 1.11 Summary of the benefits of lime-stabilized soil — 11

2. **Lime-stabilized soil as historic building fabric** — 13
   - 2.1 Traditional skills and current knowledge — 13
   - 2.2 Rammed earth — 15
   - 2.3 Earth floors — 16
   - 2.4 Clay lump, earth bricks and blocks — 17
   - 2.5 Wattle and daub — 18
   - 2.6 Earth mortars — 19
   - 2.7 Internal wall and ceiling plaster — 20
   - 2.8 External renders and screeds — 21
   - 2.9 Surface finishes — 22

3. **Building limes and hydraulic set** — 25
   - 3.1 The lime cycle — 25
   - 3.2 Hydraulic set — 27
   - 3.3 Natural hydraulic lime — 28
   - 3.4 Artificial hydraulic lime — 29
   - 3.5 Non-hydraulic lime and hydraulic set with soil — 30
   - 3.6 Pozzolans — 30
   - 3.7 Choice of materials — 32
   - 3.8 Three-stage field testing — 33
   - 3.9 Three-stage field-testing summary — 34

4.	Field testing: Stage 1 – Materials	37
	4.1 Building limes	37
	4.2 Soils	71
	4.3 Sand	81
	4.4 Pozzolans	83
	4.5 Fibres	86
	4.6 Additional materials	87
	4.7 Detrimental conditions	90
5.	Field testing: Stage 2 – Lime stabilization of soils	95
	5.1 Selection of materials	95
	5.2 Preparation of materials for trial mixes	97
	5.3 Estimation of clay content	100
	5.4 Mix proportions for stabilization	102
	5.5 Designing the final mix	104
	5.6 Selecting the form of lime	108
	5.7 Curing	115
	5.8 Field testing trial mixes	116
	5.9 Recording mix ratios	121
6.	Field testing: Stage 3 – Building elements	123
	6.1 Production methods	123
	6.2 Foundations	123
	6.3 Bricks and blocks	132
	6.4 Soil mortar	140
	6.5 Cob	142
	6.6 Rammed earth	144
	6.7 Wattle and daub	145
	6.8 Internal plaster	146
	6.9 Render and plaster test panels	149
	6.10 External render	152
	6.11 Screeds and impervious surfaces	162
	6.12 Finishes for lime-stabilized soil backgrounds	167
	6.13 Roof finishes	173
	6.14 Three-stage field testing – conclusion	174
7.	Lime-stabilized soil in civil engineering	177
	7.1 Early 20th-century applications	177
	7.2 Research and development	178
	7.3 Industrialized processes, plant and equipment	180
	7.4 Technical guidance	182
	7.5 Examples of completed projects	184
8.	Testing and national standards	187
	8.1 Current practice	187
	8.2 Field testing	188
	8.3 Lime reactivity	188
	8.4 Soil suitability	189
	8.5 Aggregate selection	189
	8.6 Pozzolans	190
	8.7 Stabilization	191

Conclusion	193
Appendices	
Appendix 1: Establishing quicklime proportions	195
Appendix 2: Mechanized equipment	197
Appendix 3: What is lime? A geological explanation	199
Appendix 4: Suitability of soils for the addition of lime	200
Appendix 5: Pozzolan testing	202
Appendix 6: Example of a test record sheet	204
Appendix 7: Compressive-strength requirements, test results and strength-gain estimates	205
Appendix 8: Field tests checklist	207
Appendix 9: Chemical formulae	209
Appendix 10: Vapour permeability: lime and lime-stabilized soil	210
Appendix 11: Earthquakes	215
Glossary	218
Bibliography	238
Index	242

Cover photos

Row 1 left. Typical Warwickshire historic timber-framed thatched house with lime-stabilized soil wattle and daub panels and limewash finish.

Row 1 right top. Comparison of magnified thin sections of lime-stabilized soil plaster from a 17th-century Devonshire manor house, with sections of new lime-stabilized soil render used during the 2015-16 Pakistan flood relief programme.

Row 1 right bottom. Lime-stabilized soil mortar to masonry of historic farm buildings in the care of the National Trust, Somerset.

Row 2. Decoration to lime-stabilized soil houses in Sindh, Southern Pakistan, as part of UKAid disaster relief and resilience programmes, indicating renewed confidence in the local building materials, now made flood and monsoon-resilient.

Row 3. The Alhambra, Grenada, Spain: 15th-century fortified palace with defensive walls and towers of compacted lime-stabilized soil.

Row 4. Examples of low-cost, lime-stabilized soil houses in Southern Pakistan – where each building element is of lime-stabilized soil: foundations, blockwork walls, sub-floors, floor surface finishes, plasters and renders, finished with limewash and traditional decoration. All carried out through community training and developed by nNGO HANDS as part of UKAid disaster relief and resilience programmes.

About the authors

Stafford Holmes and Bee Rowan are consultants in the use of traditional and sustainable building materials. They are visiting lecturers at various higher educational institutes and promote the use of natural materials for construction through educational programmes, lectures and demonstrations.

Bee Rowan is a trainer, lecturer, and consultant in natural building. She is director of UK company Strawbuild through which the lime-stabilized soil flood resilience training programmes referred to in this book have been designed and delivered, in collaboration with her colleague Stafford Holmes.

Bee has more than twenty years' experience of working in the field of sustainable construction and of designing and delivering training programmes, including accredited training in the UK and Europe, and for the international humanitarian and shelter recovery sector. Bee collaborates professionally with a broad range of associates in the areas of carbon-zero and energy-efficient buildings, specializing in the use of natural, local, and low-embodied-energy materials. Examples include an early straw bale health clinic in Mongolia for the UNDP Energy Efficiency and Poverty Alleviation Programme to a more recent partnering with Builders without Borders on Nepal's first straw bale and lime-stabilized soil house – for both seismic and monsoon resilience.

Bee is a founder member of UK and European straw bale building associations and of Re-Alliance, supporting the humanitarian and development sectors to implement regenerative change. She authored the evaluation report of the Singh government's Low Cost Housing Programme, Pakistan, and is co-author with Stafford Holmes of the practical manuals Building with Lime-Stabilized Soil for HANDS and IOM in Pakistan between 2013 and 2015, from which this publication has been developed.

Stafford Holmes is an architect. He was formerly a partner at, and is now consultant to, Rodney Melville and Partners, architects specializing in the care and repair of historic buildings. The practice promotes skills in the use of regional building materials, including building limes, for conservation and sustainable construction. He is a director of the Building Limes Development Group, a former chairman of the Building Limes Forum, and a member of the Society for the Protection of Ancient Buildings (SPAB), from which he received the Esher Award in 2017. He provides consultancy advice and expert witness to contractors and building owners when specialist knowledge of building materials and their application is required.

Stafford first carried out research and development of stabilized soils in West Africa in 1980. He is the author of *Stabilized Soil as an Appropriate Building Material*, published for the 1981 Appropriate Technology in Civil Engineering London conference. Other publications and activities include lime and lime-stabilized soil training courses in Zanzibar; reports for the Zanzibar Stone Town Conservation and Development Authority and the Intermediate Technology Development Group; lime-stabilized soil plaster and wattle and daub repairs in the UK; and the report *Evaluation of Limestone and Building Limes in Scotland* for Historic Scotland in 2003. With Bee Rowan, he compiled working manuals for building with lime-stabilized soil for HANDS and IOM in Pakistan between 2013 and 2015 from which this publication has been developed.

Stafford worked with Michael Wingate to write *Building with Lime*, first published in 1997, and is a co-author of *Lime and Other Alternative Cements* (ITDG, 1992) and *Hydraulic Lime Mortar for Stone, Brick and Block Masonry* (Donhead, 2003).

Foreword

Magnus Wolfe Murray

Massive flooding in 2010 caused unprecedented damage across Pakistan, leading to the destruction of over 1.2 million homes. This was one of the single largest natural disasters to date globally in extent of housing damage. I joined the UK Government's response – then DFID – as shelter advisor, to work alongside the UN, NGOs, iNGOs and other donors trying to support local communities and Government to design relief, recovery and reconstruction approaches that would address the immediate housing needs for over 7 million people, whilst also building resilience – so the homes of future generations could better withstand the extreme flooding events predicted for the coming years through rapid climate change.

Responsible for the design and evaluation of three disaster recovery efforts over four years in Pakistan, I was able with the support of DFID to break away from the sector's conventional re-building approaches, by utilizing lime-stabilized soil as the key flood-resilience factor in all reconstruction. Underpinning reconstruction were the training programmes in lime-stabilized soil led by Bee and Stafford, authors of this very timely book, who were absolutely mission-critical for the overall success of the recovery effort. The results of their training, mentoring and action-research are inspiring – and tangible: tens of thousands of lime-stabilized homes that have withstood consecutive years of flooding. Throughout the course of our collaboration in Pakistan, we supported communities – through exceptional partner organisations – to rebuild over 100,000 flood resilient homes, for what was little over a cost of £250 (GBP) per home. Had DFID – the principal donor during this period – followed conventional 'shelter' sector building advice, the cost per home would have been four times more, reaching 75% less families for the same investment. Or it would have required the investment of an additional £50 million (GBP) to reach the same population. I am immensely proud of this collaboration, both of the UK Government's sustained investment and confidence in us, of our implementing partners, and in Bee and Stafford's work and guidance, the critical ingredient in this ground-breaking achievement.

Those monsoon floods (or extreme rains) in the three years from 2010 to 2012 had destroyed an average of 700,000 houses per year in rural Sindh, Southern Pakistan. Subsequent and similar catastrophic events have been happening around the world in recent years from extreme climate change:

Nepal, Mozambique, South Sudan, India, Philippines, Malawi – the list goes on, and only increasing in frequency and impact.

The scale of need for so many homeless people is vast: flooding of this enormity leaves millions of people destitute each year. Recovering from such a desperate situation is a slow and difficult process. Having worked in many of these countries affected by disasters since 1990, I spend time in the communities to find out what people's priority needs are, and to try to determine what might be the most effective way to support their local recovery. Time and again I've learned that housing – or restoring people's homes so they have safe shelter – is among their top priorities, and for homes that can endure future extreme flooding events, if possible. The "build back better" concept is obvious: it's not enough to provide basic shelter and leave people to rebuild in a way that leaves them vulnerable to quite predictable flooding events in the future. As humanitarian workers, we bear a responsibility to support communities to find cost-effective solutions to build homes that can withstand predictable natural disasters, to the best possible extent.

With ever-increasing evidence that climate change is accelerating, we need to be prepared to scale up our capacity to respond with shelter recovery solutions that actually empower communities to adapt and cope, now, and for generations to come.

So the question we need to pose to the humanitarian community and local Governments is: how best can we meet this challenge with the limited resources and funds available? How can we reach as many people as possible while maintaining a high standard of construction so that the new homes will not collapse next time there is a fierce rainfall/heavy monsoon, hurricane, cyclone or flood that leaves water standing across entire landscapes for months on end? And how can this be done in an environmentally sensitive way, so minimizing the carbon loading of such large construction programmes?

We need to recognize that most people live in low-cost homes they have constructed themselves, often with local materials such as wood, stone and earth. In most poor, rural communities in developing countries, people don't have access to insurance or affordable finance. To avoid sinking into debilitating debt, these people need ways to rebuild in their local and traditional (vernacular) way with minor, low-cost improvements so their homes are more resilient to extreme flooding.

Those who do not learn from mistakes of the past are bound to repeat them. Examples of historic buildings that have stood the test of time are numerous worldwide, as UNESCO World Heritage will testify. Lessons can be learnt from the traditional architectural vernacular and from cultures that have survived for centuries. Investigations, initiated by the desire to conserve and repair historic buildings have led to much of the scientific evidence, understanding and application of materials that will last well through adverse weather conditions. This technology is not only most helpful for the sympathetic repair and conservation of historic building fabric worldwide, but also for disaster relief and recovery, particularly in remote rural areas of the world.

This book depicts the strategy – of which the key ingredients, widely available, are limestone and soils, particularly those with a clay content. When limestone is burned and combined with local clay soil, and/or pozzolans, (locally available burnt brick dust or various types of ash), it can create a range of low cost, flood resilient and strong building components: foundations, floors, wall blocks, renders, plasters and roof screeds that remain stable in wet conditions and under water. Field tests demonstrate many of these mixes are stable under water for many years, whilst also continuing to increase in compressive strength.

A critical aspect to understand is to respect and work with local and traditional building practices, designs and materials, rather than impose external notions of what is appropriate. Architectural traditions demonstrate that local, natural materials of earth for walling (whether of adobe, cob, rammed earth or wattle and daub) and of thatch for roofing, can be the predominant vernacular of a region where communities will be familiar with, and skilled in, building with these materials. Although earth is an entirely appropriate material with proven longevity in dry conditions, the vulnerability of earth buildings when exposed to water for any length of time is clear, in how quickly they erode or dissolve. The lack of extended roof eaves in many cases also allows rain to enter the top of the walls leading to wall collapse. This is why training and up-skilling on how to effectively use locally available lime to stabilize building materials, whilst still utilizing and valuing traditional building designs and methods, is so vital and can have so much meaningful impact. Communities do not always need external contractors or builders to construct their housing for them, and certainly not of expensive and environmentally damaging, modern conventional materials.

So, here it is: a guide to building and repairing low cost, low carbon, durable, monsoon and flood resilient buildings for the increasing extent of areas vulnerable to flood events across the world.

Magnus Wolfe Murray
Humanitarian/Reconstruction Advisor
UK Department of International Development (DFID); Pakistan: 2010–2014, South Sudan 2015-2016; Nepal: 2016–2019
International Organisation of Migration (IOM) Mozambique: 2019–2020

Preface

The focus of this book is to describe methods by which the research, development and field testing of lime, soil and other local materials may benefit repairs to historic buildings, assist with sustainable construction, and support lasting disaster recovery and community resilience.

The links between the use of lime-stabilized soil in historic buildings, its suitability for disaster resilient design and national standards for civil engineering are explored in Chapters 1 and 2. Lime-stabilized soil is one of the most sustainable and ecologically friendly building materials that is generally readily available. It has been used in historic buildings for over 700 years, is used successfully in disaster recovery and resilience programmes, and is and has been used for road building and other civil engineering applications worldwide. Chapter 3 provides a scientific overview of the various applications of lime in building, both historically and today, in many parts of the world.

There are many mass-produced water-resistant building materials, but these can harm traditional and sustainable buildings. Their manufacture may also cause ecological damage. Technically advanced materials are not always available or appropriate for many communities, mostly due to difficulties of transport and cost. Chapters 4, 5 and 6 describe methods of low-cost field testing using materials that will have greater durability in persistently wet conditions, and under flooding, than many alternatives currently employed.

These chapters on low-cost materials testing, preparation, application and aftercare in the field are intended for managers and practitioners in disaster and flood-resilience training programmes who wish to instruct builders in these techniques. They are also an important resource for anyone embarking on using lime-stabilized soil for the repair of historic buildings or in new construction projects.

Acknowledgements

Acknowledgement and thanks are given here to Practical Action Publishing, previously the Intermediate Technology Development Group (ITDG) and Intermediate Technology Publications, who are responsible for the supporting publications by John Norton, author of *Building with Earth,* and Michael Wingate, co-author with Stafford Holmes of *Building with Lime.* With permission, some sections from these publications have been adapted and incorporated into this book as they provided excellent source material for describing ways in which stabilized-soil construction methods may be used in conservation and to assist rural communities. *Earth Construction* by Hugo Huben and Hubert Guillaud, from the same publisher, has also been most helpful. Thanks are also due to the European Leonardo funded ECVET training programme materials in Earth Building and Clay Plasters, in which Bee Rowan was a co-creator. Details of sources of information can be found in the bibliography.

We are grateful for the support given by Magnus Wolfe Murray, Humanitarian Shelter and Reconstruction Advisor to DFID (UK Aid), and to HANDS Pakistan, for their vision, commitment and assistance with the flood-resilience work, training programmes and first iteration of this book: edition 1 of the lime-stabilized soil training manual, particularly to Dr Shaikh Tanveer Ahmed, Chief Executive, and Mustafa Ghulam-Zaor, Services Chief Executive. We are also grateful to IOM for their input and support during the extended flood-resilience programmes and for the second, extended edition of the manual. Very special thanks to HANDS Shelter Programme Manager Qazafi Memon and District Project Manager Hakeem Darejo, who became the Strawbuild Pakistan (and later, Nepal) field team, without whom the great results of this work would not have been possible.

Thanks to all those committed trainers and motivators introducing lime to villagers where the skills for using it are most needed, and not least to the villagers themselves, many of whom have lost almost everything, often repeatedly. They are embracing lime-stabilized soil construction methods, and may become a model for the wider world as they rediscover this ancient and effective method of using local soils and lime, the benefits of which range from conservation and maintenance of historic buildings to repairs and new build.

The technical assistance, good will and considerable help given by all at Rodney Melville and Partners is greatly appreciated, especially that of Maureen Burford for her generosity, exceptional patience and enduring secretarial help.

Finally, very special thanks to Juliet Breese, our wonderful illustrator, whose delightful and highly skilled work is central to helping make this technology accessible to many.

Abbreviations

AASHTO	American Association of State Highway and Transportation Officials
ARBA	American Road Builders Association
ASTM	American Society for Testing and Materials
BLA	British Lime Association
BLF	Building Limes Forum
BRE	Building Research Establishment (UK)
BS	British Standard
CBR	California bearing ratio
CEN	European Committee for Standardization
CIOB	Chartered Institute of Building
DFID	Department for International Development (UK)
DoT	Department of Transport (UK)
DPM	Damp-proof membrane
DRR	Disaster risk reduction
ECVET	European Credit System and Certification in Vocational Education and Training
EN	European Standard
FAA	Federal Aviation Administration (USA)
GGBS	Ground granulated blast-furnace slag
GLC	Greater London Council
HANDS	Health and Nutrition Development Society (national NGO, Pakistan)
HRB	Hydraulic road binder
iNGO	International non-governmental organization
IOM	International Organization for Migration
IP	Implementing partners
IPCC	Intergovernmental Panel on Climate Change
IS	Indian Standard
ITDG	Intermediate Technology Development Group
KRI	Kurdish Region of Iraq (Northern Iraq)
LSS	Lime-stabilized soil
MCV	Moisture condition value
NED	Nadirshaw Eduljee Dinshaw University of Engineering and Technology (Pakistan)
NGO	Non-governmental organization
NLA	National Lime Association (USA)

nNGO	National non-governmental organization
NZ	New Zealand
OMC	Optimum moisture content
OPC	Ordinary Portland cement
PFA	Pulverized fuel ash (fly ash)
ppm	Parts per million
psi	Pounds per square inch
SHB	Soil with hydraulic binder (pozzolan)
SPAB	Society for the Protection of Ancient Buildings (UK)
STCDA	Stone Town Conservation and Development Authority (Zanzibar)
UCS	Unconfined compressive strength
UNCHS	United Nations Centre for Human Settlements
VOC	Volatile organic compound

CHAPTER 1
Lime-stabilized soil: an invaluable resource

1.1 Introduction: economic and environmental benefits

This book has been prepared as a guide for using lime-stabilized soil (sometimes abbreviated to LSS), in both repairs to historic buildings and in new construction. This is of relevance in many places throughout the world, predominantly rural areas and those where severe flooding has had a disastrous effect on un-stabilized earth structures, such as in recent years where hundreds of thousands of houses have dissolved in standing floodwater. While there are now many industrially manufactured water-resistant materials, these often harm traditional building materials and can cause ecological damage. In addition, the associated difficulties of transport and cost mean technically advanced materials are not always available to, or appropriate for, many communities. However, results from two flood-relief programmes in Pakistan (2012 to 2015 disaster recovery and resilience), and more recent programmes responding to the devastating 2015 earthquake in Nepal, have demonstrated the value of developing lime-stabilized mixes for a range of building components.

On-going tests show that many of these mixes, all of local soil stabilized with only small amounts of lime, remain stable under water for several months – many for over two years so far. Wet compressive-strength tests indicate that many cured mixes are strong enough to be used in the construction of new two-storey buildings. These results are currently being correlated with those from laboratory tests, but the successful field tests are compelling and are supported by independent laboratory testing of lime-stabilized soil for Arup Consulting Engineers London, described in their 2017 report *Flood Resilient Shelter in Pakistan – Evidence Based Research*.

Having seen the results of the flood-relief programmes, village communities are continuing to use these methods. Some of the most vulnerable have built tens of thousands of new, lime-stabilized soil houses for themselves and their families after being trained through LSS flood-resilience programmes funded by the UK Department for International Development (DFID). There is already evidence of growing local entrepreneurship: villagers who have recognized the value of low-cost flood-resilient material are selling lime-stabilized soil building blocks and tested, good-quality lime products. The potential for lime to be used in this way worldwide is reflected in beautiful historic buildings that have demonstrated durability and withstood repeated floods. Lime has been used for thousands of years and is culturally an essential material in many traditional and historically important buildings. Building limes have been used predominantly for lime mortars, plasters, renders, and decorative work. The techniques for mixing lime and other modifying materials with earth to form a stabilized material for building elements including foundations, floors, walls and finishes have until recently been a forgotten art, which is now in the process of rediscovery and renewal.

One of the economic benefits of using lime-stabilized soil instead of the frequently used unfired 'earth' – i.e. un-stabilized sun-dried soil blocks, cob walls, and other elements which are often only modified with cow dung, and/or natural fibre – is the dramatic difference between not having to rebuild, particularly following floods, or having to rebuild after saturation and collapse. A 'stable' mixture will not dissolve and return to mud when under water. A basic fact, not widely understood at present, is that many soils can be stabilized with the addition of as little as 3–10% of well-prepared lime, provided the soil is of the correct composition. This has been confirmed by laboratory tests in various countries, not least the USA, where lime-stabilized soil has been increasingly used for road sub-bases and canals.

Buildings made from lime-stabilized soil do not have to be rebuilt after each flood, and lime-stabilized soil renders and other elements need little or no renewal or maintenance after heavy rains, flood or water ingress. The cost and labour for simple field tests and production of lime-stabilized soil are significantly less than the cost of rebuilding after floods or, in terms of maintenance, the cost of renewing render and other exposed areas after heavy or even moderate exposure to the elements. The cost of constructing a lime-stabilized soil building occurs once, whilst renewal of un-stabilized earth may have to be repeated many times. Another significant economic benefit is the dramatic cost saving made when using lime-stabilized soil rather than conventional building materials. Completed projects across a wide geographical spread have demonstrated this to be as much as 70%, as detailed in Chapter 3. There are also substantial savings to the carbon footprint of any construction programme: as the lime cycle (Figure 3.1) demonstrates, pure lime can reabsorb carbon dioxide (CO_2) and, if fuel is excluded, this makes it almost carbon-neutral, whereas cement and fired-brick manufacture have a very high carbon footprint. Further low-impact advantages of using lime in preference to more complex building materials are evident when comparing environmental pollution during manufacture and whole-life costing.

The principal production process for both cement and lime is burning limestone, and in many countries this is done locally, on a small scale. At present, lime for various uses is manufactured in kilns that burn a variety of timber and vegetation for fuel, although coal, gas, oil or electricity may also be used. In the short term, the disadvantages of small-scale wood burning can be mitigated by developing fuel-wood plantations. These may be of dual-purpose tree species selected for functions other than only fuel, such as food crops. Plantations like these would create additional benefits, not just timber for fuel, and would reduce overall emissions from lime production by reabsorbing carbon dioxide during regrowth. Solar, hydroelectric, and geothermally powered kilns that would not produce carbon dioxide need development.

Where ambient temperatures are high, the choice of dense concrete and other climatically inappropriate materials will often be detrimental to the internal environment of a building because they raise and lower temperatures to well outside the comfort zone. Soil blocks have lower thermal conductivity than most building materials, other than lightweight insulation blocks. Laboratory tests and field monitoring have shown that soil construction methods lead to much improvement in the indoor environment, with reduced indoor air temperature fluctuations over a 24-hour period (Fathy, 1986). Energy requirements and costs for the cooling and heating of internal spaces are thereby reduced.

Mix designs for specific building elements benefit from the introduction of other materials selected for their compatibility and reactivity with soil and lime. Associated materials for use in conjunction with the lime stabilization of soil, including brick dust and various types of ash used as pozzolans, are often locally available in large quantities and may well be low-cost waste or by-products. In many cases, the performance of an element may be enhanced by modifying mixes with particular types and quantities of fibres and aggregates, subject to the soil's composition and mineralogy.

An understanding of how to select good-quality lime is an important part of the process. This helps users to avoid lime that may not have had adequate preparation, transportation, or storage. Ensuring that good-quality lime is selected, and a satisfactory standard maintained, is therefore a key component of the first stage of testing detailed here. Quality control is cost-effective, because it helps to ensure the avoidance of defective lime (which is often sold at the same price), saving duplication of effort and additional cost, should poor-quality materials cause failure and result in the need to rebuild.

This book is not intended to be a construction manual. The aim is to describe methods of field testing and developing local materials that may benefit appropriate repairs to historic buildings, assist construction in rural communities, support disaster risk reduction and recovery, and offer sustainable, durable and ecologically benign materials for buildings that support climate-change mitigation.

Simple but thorough field tests for all materials used, complete soil stabilization, and appropriate modification of the mix for each building element have been central to past achievements, and detailed descriptions of these form the core of this work.

1.2 History and durability

Soil structures have a long history and were one of the earliest means of creating shelter. Soil construction techniques were almost certainly well developed in many parts of the world before the discovery of building limes. Historic buildings incorporating soil and lime in their construction and finishes are testament to the durability of these materials. Local soils and building limes were not only used individually, but could also be combined in different ways due to their compatibility. Limewash and lime plasters, for example, are suitable finishes for compacted-earth walling and daub panels. Many buildings constructed with these materials remain standing today after centuries of continuous use.

The abundance, availability, and compatibility of various soils and building limes provided opportunities for their combination and development, which resulted in durable and well-finished structures. As well as applying lime-based finishes to soil surfaces, it was found that the addition of lime in appropriate proportions to various soils that had a clay content could produce the dramatic change from a soil that dissolves in water to one that does not. It is probable that this was originally discovered by trial and error rather than through an understanding of the science.

Examples of historic buildings that demonstrate the durability of lime-stabilized soil include the following:

- The Alhambra fortified palace, Spain, built during the 13th and 14th centuries, has defensive walls built of compacted soil, gravel conglomerate, and lime (Salmerón, 2007). (See photograph 1c.)
- 16th-century forts in East Africa have mortars and renders of lime and lateritic soil mixes, with walls and internal surfaces that have remained intact for over 400 years. (See photographs 1a and 1b.)
- Domestic buildings in England of soil and lime, dating from the 16th century onwards, continue to be maintained and occupied to the present day. These include timber-framed houses with lime-stabilized wattle and daub panels in the Midlands, lime-stabilized earth mortars in Somerset, and lime-stabilized soil plaster and render on compacted cob in Devon. (See photograph 3a.)
- One of the largest historic forts in Pakistan, Kot Diji, Sindh, built in the 18th century, is constructed of burnt brick with soil and lime mortars, plasters and renders, and these materials appear to be included in its large water tanks.
- 19th- and 20th-century domestic stone buildings and war-damaged higher-status houses and palaces in KRI, Northern Iraq, have mortar of 'red soil and locally burnt lime', as identified by Dohuk Antiquities and Heritage Department Archaeologist and lime producer Ehsan Kovan.
- Temples of earlier date in Nepal currently being repaired and conserved following earthquake damage have internal and external burnt-brick walls with earth mortars, some finished with lime-stabilized soil plasters and renders. (See photograph 16c.)

Archive research by Nigel Copsey (2016) has identified historic written records of lime and earth mixes used in the British Isles, including the following:

- *Building Accounts* of All Souls College, Oxford (1438–1443), which records the purchase of '289 quarters of quicklime' plus 'red earth with sand for making mortar' (1 quarter = 291 litres).

- *A Dictionary of Husbandry, Gardening, Trade, Commerce and All Sorts of Country Affairs: Volume I,* by Bailey and Worlidge (1726), describes how to make mortar: 'Take loam or a brick-earth and mixing therewith some good lime, temper them very high till they become tough, smooth and gluey [...] Afterwards temper some of the earth that the wall was made of, with a little more lime than was used for the wall, which you must be sure to temper very well, and with the mortar, plaster all your wall well on the outer side which will keep off the weather.'
- An 1841 farmers' guide for the use of small farmers and 'cotter tenantry' in Ireland advises that 'floors may be of earth or clay, well mixed with sand and lime, and beaten hard and smooth and raised from eight inches to a foot above the level of the ground outside'.

Laboratory analysis by Revie (Construction Materials Consultants Ltd, personal communication, 2014) of an earth–lime plaster from the staircase walls of the 17th-century York House in Malton, indicates that the original ratio was approximately one part lime putty to eight parts soil aggregate or one part quicklime to sixteen parts soil.

1.3 Ecological advantages

There are many good reasons for choosing lime as a binder and stabilizer other than it being an excellent material for stabilizing clay soils.

There are abundant limestone resources worldwide, so lime can often be produced and used locally. Small-scale lime production can therefore help to reduce transport distances and associated pollution, improve sustainability, and benefit communities in rural areas, resulting in ownership of local, durable and sustainable building methods for climate change mitigation and adaptation for resilience to floods and monsoons.

Various recently developed building materials, mass-produced using advanced technology, have become a significant cause of pollution with emissions into air, land, and water increasing at an alarming rate. The Intergovernmental Panel on Climate Change special report (IPCC, 2019) stresses the urgent need to limit planetary warming to less than 1.5°C above pre-industrial levels as agreed by governments in Paris in 2015. This target is directly linked to the minimizing of harmful carbon emissions and to climate change mitigation measures at international, national and local levels. As recently declared by the UN Climate Secretariat (June 2019) "There is (already) universal agreement on immediate rapid global emissions decline to avoid total planetary catastrophe – it is the bottom line for our survival." Where practical, the use of lime and lime-stabilized soil, in place of materials such as Portland cement that produce high unrecovered, carbon emission levels during manufacture, can play a significant role in emission reduction.

Pollution is also caused by many manufacturing methods in and for the building industry, often synthetic, which produce toxic waste, dust or fumes, ozone-depleting chemicals and acid rain as well as carbon dioxide. Installing some of these materials in a building have also been shown by various studies to be detrimental to health. Asbestos is a well-known example, as well as off-gassing from such as added formaldehyde glues and volatile organic compounds (VOCs) in common building products. As soil is not manufactured, this cause of pollution to either the environment or human health is eliminated. In addition, one of the many properties of clay is to absorb rather than emit pollutants.

Lime does emit carbon dioxide during its manufacture but the majority is reabsorbed as it sets, and the proportion of lime required to stabilize suitable soils is small, generally around 3% to 10%. Lime-stabilized soil is therefore predominantly soil – 90% or more of each mix is soil. Small-scale lime production can often be local, and a wide variety of fuels may be used for lime burning, which could be from sustainable fuel sources.

1.4 Health and sustainability

One of the most important benefits of lime is the contribution it can make to improving human health and hygiene. Lime can be used, for example, in the purification of water. Lime's mild disinfectant and antiseptic qualities are due to its high alkalinity, which is also why it can be used to improve soil for agriculture. The materials with which a building is constructed have in recent years been shown to affect the health of its occupants to a greater degree than was generally realized. Lime mortars, plasters, and renders, including those used in conjunction with earth construction, are more vapour permeable (able to breathe – see Appendix 10) than denser materials, many of which have been rapidly developed over the last century and are impermeable. The consequent 'sealing' of interiors with impermeable materials is a main contributor to the on-going epidemic of 'sick building syndrome', which places the health of the occupants at risk. The long-term disadvantages of using toxic materials and introducing impermeability into the building fabric have been increasingly recognized, especially as the selection, manufacture, and installation of building materials may have serious consequences for the environment. Materials best avoided, or kept to a minimum, are those that necessitate high energy use in their manufacture, produce dangerous levels of carbon dioxide and toxic waste, and are not recyclable. This has been thoroughly documented in the *Green Building Handbook* (Woolley et al., 2016).

Softer, more permeable materials provide a comfortable environment. Moisture is allowed to evaporate and is not trapped, as often happens when impermeable materials are used. Trapped moisture can cause decay in adjacent materials and lead to mould growth, rot and infestation by providing a breeding ground for fungi and insects. Permeable materials allow the evaporation of moisture and so help protect adjacent built-in materials, such as timber and ferrous metals, from damp conditions and associated decay. Building limes facilitate the use of other soft and vapour-permeable building materials due to their compatibility with lime. These materials, frequently used in historic buildings, may well be both low-cost and locally available. Soft brick, cob, earth blocks, wood, wattle and daub, thatch, straw and hemp are examples of these.

One of the evident ecological benefits of lime is its contribution to environmental sustainability. Efficient, small-scale local lime production can result in lime binders having significantly less embodied energy than more complex binders such as Portland cement, thanks to the reabsorption of carbon dioxide in its setting process and shorter transport distances. Pure-lime production can be made almost carbon-neutral as there are many ways to provide the energy needed for burning. Magnus Wolfe Murray, humanitarian shelter advisor for DFID, has shown that using lime-stabilized soil rather than conventional and energy-intensive fired brick and cement for the construction of 50,000 new-build houses can save around 180,000 tonnes of carbon dioxide (based on energy calculations from the University of Bath, UK), a reduction of 70% (Table 1.1). Developing solar-powered and bio-gas kilns and fuel-wood plantations in conjunction with small-scale lime production in rural areas, as part of a holistic and sustainable approach to the use of lime, would further enhance its ecological benefits.

Table 1.1 Possible CO_2 savings from the use of lime-stabilized soil rather than conventional and energy-intensive fired brick and cement for the construction of 50,000 houses. This table shows that by using lime-based mixes and materials for this number of houses there is a possible reduction of almost 180,000 tonnes of CO_2 in the production process.

Item	Quantity per house	CO_2 emissions from material (kg)	CO_2 emissions per house (tonnes)	CO_2 emissions per 50,000 houses (tonnes)
Fired bricks	5,500 bricks	0.55 per brick (0.23 per kg)	3.03	151,250
Cement (Average CEM I Portland cement, 94% clinker)	600 kg	0.95 per kg	0.57	28,500
Lime replacement	50 kg	0.234 per kg	0.0117	585

Source: Wolfe Murray (2014).

There are various other additional benefits to using lime in building construction. Lime is made of fine particles and is a sticky material, which can bind firmly and gently to other surfaces, providing early adhesion. It also has good workability. Over the longer term, the precipitation of free lime means it can be 'self-healing' and so repair fine cracks.

The combined benefits of the low cost of lime, clay soil, and pozzolans, the ecological and health advantages of their use, the fact that limestone is often

8 BUILDING WITH LIME-STABILIZED SOIL

locally available, and that building limes may be produced on a small scale are all important considerations.

In this context, incorporating lime is one of the most appropriate methods of stabilizing soils for building elements that require a binder for their modification, improvement, strength gain and resistance to water, especially given our current environmental fragility.

To summarize, some advantages of using lime are that it:

- has low embodied energy
- is carbon-neutral (subject to the burning method)
- has healthy, hygienic, antiseptic qualities
- assists the evaporation of moisture
- is compatible with other natural and traditional materials and suitable for historic building and earth fabric repairs
- protects other materials
- improves indoor comfort
- has good workability
- enables beautiful finishes
- can be produced locally on a small scale
- can be low-cost
- has a proven track record of durability over thousands of years
- stabilizes soil
- assists sustainable development, encourages skills development, and supports self-sufficiency
- assists flood mitigation.

1.5 Low-cost construction

One way of achieving economies in the building industry is by maximizing the use of local resources, including employing and developing the skills of local people. Creating greater self-sufficiency increases economic resilience and reduces or eliminates reliance on distant materials, skills and external funding for construction.

Lime-stabilized soil is compatible with other sustainable building materials which may also be available or developed locally. Learning from vernacular building traditions and training local people where required, in the appropriate use of the many various materials including earth, provides an opportunity not only to rebuild cost-effectively, but also to assist communities become more self-sufficient and resilient.

Comparative costings of lime-stabilized soil construction materials with burnt brick and cement/sand mortars and plasters undertaken in Southern Pakistan (2015–2016) demonstrated an average saving of about 70% across different geographical regions, calculated by different organizations.

Since the price of local soil is negligible, the economic case for reducing the cost of transport, imported materials and skills and enhancing local development instead is self-evident.

1.6 Diversity of building elements

Lime-stabilized soil mixes and their modification provide a versatile base for a wide range of flood-, monsoon- and water-resilient building elements. This book recommends a three-stage field-testing programme to select suitable materials and determine appropriate proportions of all materials to develop various mixes suitable for foundations, floor slabs, screeds, bricks, blocks, mortar, render, and plaster. As a combination of clay and lime is the binder, and lime is the catalyst for both carbonation and hydraulic set, a similar curing and aftercare regime applies for building with all types of lime. Field-test results of completed mix samples across several programmes in Pakistan and Nepal have shown that it is possible to create materials with compressive strengths of over 4.75 N/mm^2 (700 psi). These values were obtained by the penetrometer testing of samples that had been submerged continuously under water, some for over a year. In 2017, independent laboratory test results for Arup Engineering, undertaken by the New Material Testing Facility at the NED University of Engineering and Technology, showed that lime-stabilized soil blocks made during the programme achieved 7 N/mm^2 (1,015 psi). The same series of tests showed the cement-stabilized soil achieved significantly less, only up to 1.5 N/mm^2 (Arup, 2017).

1.7 Disaster recovery and rural development

Natural disasters including floods, hurricanes, cyclones, earthquakes and forest fires occur worldwide and do not respect political boundaries. As well as affecting cities and towns, they can have devastating effects on rural areas.

It is difficult to reach remote villages and rural communities in many parts of the world, particularly in times of crisis. Transporting building materials over long distances is slow, costly, and seldom practical. If 90% or more of the bulk building materials are on or close to the construction site, the difficulties and costs of transport are eliminated.

Low-cost field testing and simple construction methods with lime-stabilized soil are suitable for both community- and self-builds, and for larger disaster-recovery building programmes. They are also well suited to the practical and sympathetic conservation of historic buildings where repairing like with like is important.

1.8 Simple field-testing methods

In order for a construction method to be reliable, all materials used should be of an approved standard and preferably of the best quality. When using lime-stabilized soil, there are simple ways to test the materials and mixes on site. There is no need for laboratory conditions and elaborate and expensive equipment, although subsequent laboratory testing to verify field-test results is advisable.

Disaster-recovery programmes which have little or no funding and are operating in remote areas cannot afford complex test facilities and equipment, but need ways to ensure all materials and mixes are of a satisfactory and reliable standard. The composition of soils is variable and not all are suitable for stabilization, so it is important to establish the type of local soils and other materials available and ensure that they are able to produce satisfactory mixes. There are methods of testing the materials, mixes, and lime-stabilized soil building components using simple, low-cost field equipment.

It is recommended that field tests are used firstly to test individual materials: for example, lime and pozzolans for reactivity, sands for particle-size distribution, and soils for clay content. Secondly, lime-stabilized soil mixes should be tested for adequate stabilization, strength, and durability in wet conditions. Thirdly, modified mixes required for individual building components need to be tested as appropriate, and throughout the manufacturing process. Examples of particular tests include those for establishing and maintaining the best proportions of fibres and/or sand in a plaster; lime types; aggregate sizes and compaction required for blocks; and the permeability of render. Durability and strength should also be tested, as these may increase (or reduce) over time depending on soil type, mix ratios, and environmental conditions.

Simple tools, containers and sealants make these tests cost-effective. Most field-test equipment can be made on site or obtained close to it (see 4.1.2).

1.9 Established and current use

Some established uses for lime-stabilized soil have been described in *Building with Lime* (Holmes and Wingate, 2002). These include the following:

- *Block strengthening* – 'Properties of bricks and blocks' in Building Research Establishment (BRE), *Overseas Building Notes* Nos. 154 and 184 (1980) confirmed that compacted soil blocks stabilized with 6% lime in Ghana achieved a compressive strength of $4.5\,N/mm^2$ (660 psi) at 28 days.
- *Block production* – Norton describes the production of lime-stabilized soil compressed blocks in his publication *Building with Earth* (1997).
- *Render and plaster finishes* – A render mix of lateritic earth, sand, and lime in the ratio of 6:2:1 is used in Natal, South Africa.
- *Road construction* – Lime stabilization is used and evaluated in soil-lime in the USA to a level suitable for the construction of modern road bases; similarly, the UK's Department of Transport (DoT) includes lime stabilization of subgrades in its specifications (Department of Transport Highways Agency, 1986).
- *Stabilizing waterlogged ground* – The use of quicklime on waterlogged ground is described in a BACMI advertisement and description of the lime-stabilization process (Buxton Lime Industries, 1990).

1.10 National standards for lime-stabilized soil

A number of countries currently have national standards for civil-engineering applications of stabilized soil. These include the American Society for Testing and Materials (ASTM) standards in the USA and the European Committee for Standardization (CEN) standards for most European countries, including the UK. Although intended for substructure and civil-engineering applications, the principles are similar for building elements and superstructure. The basic lime-stabilization process, however, needs modification to make it suitable for a range of building components such as blocks, mortars, plasters, renders, and screeds, which are not covered by these standards but are the focus of following chapters.

1.11 Summary of the benefits of lime-stabilized soil

There is an extraordinary link between civil engineering, the repair of historic buildings and low-cost construction for rural communities, especially for disaster risk reduction and recovery. It is lime-stabilized soil. As a climate adaptation and disaster-mitigation tool, understanding how to use lime to stabilize local soils offers empowerment to flood-risk communities who cannot afford expensive alternatives. Technically advanced materials are frequently less appropriate or effective than lime-stabilized soil, and may well also have higher carbon, health, and environmental costs.

To summarize, the key benefits of building with lime-stabilized soil include the following:

- Fully lime-stabilized soil does not dissolve under water.
- It is stable in wet conditions for building construction above ground level, as well as for substructure and ground works.
- Lime-stabilized soil offers greater ecological benefits than many conventional building materials and can be carbon-neutral, or close to it.
- There are economic advantages to lime-stabilized soil that enable low-cost construction.
- Lime-stabilized soil is suitable for a wide range of building elements and finishes.
- Lime-stabilized soil is an appropriate technology for disaster recovery and rural development.
- Simple, low-cost field-testing and quality-control methods have been developed that allow reliable and reproducible local construction.
- The long-term durability of this material has been confirmed by both historic and current use, examples of which are given in this book and are summarized in Chapters 9 and 10 of *Building with Lime* (Holmes and Wingate, 2002).
- Lime-stabilized soil can and has been successfully used for civil-engineering purposes, particularly the construction of roads, runways,

canals and dams, for which it has been used extensively throughout industrialized regions of the world (Chapter 7).
- Lime-stabilized soil is compatible with, and assists the use of, other ecologically sustainable construction methods and building materials.
- Well-prepared, compacted lime-stabilized soil can achieve high wet compressive strengths which exceed the requirements of the *London Building By-laws 1972* (GLC, 1973) for walls of two-storey buildings (see Appendix 7).

CHAPTER 2
Lime-stabilized soil as historic building fabric

2.1 Traditional skills and current knowledge

The recent rediscovery and increasing understanding of building limes in the UK is testament to the benefits of reviving traditional materials and skills. The disappearance of this knowledge in one generation, however, demonstrates the speed with which building skills and expertise can be lost.

Building materials and traditional methods once commonly accepted may be neglected, following developments in the conventional building industry with the intention of improvement through change. The extent of this is increased by the introduction of new materials that can be mass-produced on an industrial scale. These materials may initially appear to be superior to the original, and disadvantages may not be apparent at first. The current lime revival, initiated in connection with historic building conservation and now extending to new build, demonstrates the value of retaining both knowledge and skills that have been successfully used in the past.

The importance of like-for-like materials in the context of repairs to historic buildings has led to increasing research into the earth mortars, plasters, and renders which are integral to many existing buildings.

Historically, it was common practice to use lime in conjunction with earth, as these two materials are compatible and, when used well together, produce building materials with improved durability. Examples include lime pointing over earth bedding mortars, lime skim coats or limewash over earth plaster or wattle and daub; lime renders on cob, compacted-earth walling, and lime-stabilized soil render. (See photo 2.)

More examples of how lime-stabilized soil has been used in historic buildings are being recognized and examined by those concerned with true conservation. For example, recent analysis of lime and lateritic soil mortars from a 19th-century building in Zanzibar concluded that the mix contained between 5% and 10% lime (3D design-development-display OEG, 1992). (See photo 5a.)

Recognizing that a soil has been lime-stabilized is not always straightforward. One of the more obvious indications that lime has been incorporated in an earth mix is the presence of small but clearly visible lime inclusions, usually carbonated lime granules. These may be from late-slaking particles in quicklime, or from a breakdown of the carbonated crust on (stored and neglected) putty. Late-slaking lumps of (un-sieved) dry hydrate may give a similar appearance. However, it is possible that such granules may be white aggregate particles and not lime at all.

Experience has confirmed that clay soils may be stabilized with lime putty, not just quicklime. Quicklime is an effective stabilizer and works quickly, as the name implies, but it needs to be of the best quality and in the form of very small lumps, granules or fine powder. If not, late slaking causes problems and can destroy finer work, such as plaster and render finishes. Plasterers working in recent flood-relief programmes favoured lime putty for stabilizing earth plasters and renders, and lime in the form of putty was probably applied to earth renders in the past. It is difficult to detect well-distributed good-quality putty in lime-stabilized soil with the naked eye, due to the minute particle sizes involved. Critical reactions take place with clay particles less than 0.002 mm (2 microns) across. A good-quality lime putty, well mixed in the proportions of 1 part putty to between 10 and 20 parts earth, may not be visible in the field, although it may be observed with laboratory equipment. A simple field test is to place a sample under water: most un-stabilized soil will dissolve 20 minutes to half an hour after immersion, if not sooner, but fully stabilized soil will not. An alternative test procedure is to determine the pH of the soil. ASTM standards advise that sufficient lime should be added to the soil to achieve a pH of not less than 12.4. If the pH is lower than this, the soil may not be fully stabilized. (See photo 13.)

Long-term durability is an important factor. Well-stabilized soil will remain firm and not dissolve after being submerged in water for over a year – for over two years, in some recent tests. In a 17th-century English manor house that caught fire, lime-stabilized soil wall and ceiling plaster remained stable after being saturated with water from firemen's hoses for several hours (Historic England, 2015). Laboratory analysis confirmed that lime, probably in putty form, had been incorporated in the original earth plaster mix, although when inspected on site it was not visible. (See photos 2 and 19.)

The way in which lime combines with and consolidates sand for lime–sand mixes involves different chemical reactions to those in which lime stabilizes soil. Lime–sand mortars (commonly known as lime mortars) consolidate and set because lime hydrate, $Ca(OH)_2$ (calcium hydroxide), surrounding the sand particles becomes solid carbonated lime, $CaCO_3$ (calcium carbonate). The process for hydraulic lime is similar but here there is also a chemical set involving primarily calcium silicates and calcium aluminates. Lime–sand only mixes require a substantial proportion of lime to sand – in the order of 1:3 to 1:1 (that is, from 25% to 50% lime in the mix).

The lime stabilization of a clay-content soil, in contrast, requires the addition of only between 3% and 10% lime depending on the quantity and type of clay – substantially less lime than is required for lime–sand only mixes. The amount of lime will also vary with the type of soil, and whether quicklime or lime putty is used. However, there is still a significant reduction in the proportion of lime to aggregate and this offers both ecological and cost benefits.

In *Earth Construction* (1989), Houben and Guillaud advise that soil may be 'waterproofed' (or made water resistant) by the addition of potassium sulphate, K_2SO_4. Much organic matter contains potassium and sodium, and there is a long tradition of adding animal and plant products such as cow dung and oils or fats to lime mortar and soil mixes. Unfortunately, the precise origins of organic additives used in historic mixes are difficult to trace, even with advanced laboratory equipment, but if they are present, investigating local traditions and customs may give an indication of their origin.

The following sections of this chapter give examples of how lime-stabilized soil has been used over the centuries for building elements, from rammed-earth walling to surface finishes, and to protect and modify earth buildings and structures in various parts of the world. Chapters 3, 4 and 5 describe a three-stage field-testing method for lime-stabilized soil construction covering most building elements. These sections include material from earlier training manuals, revised and developed by the authors and based on further practical experience.

2.2 Rammed earth

Whilst it is clear that rammed-earth structures have been built in many countries and probably well before recorded history, written evidence that lime was used as a stabilizer is scarce. One reason for this may be that the relatively small proportion of lime required for stabilization and the importance of thoroughly incorporating it in the mix make the visual distinction between lime-stabilized and un-stabilized compacted soil (or rammed earth) difficult. A considerable number of lime-stabilized soil buildings may therefore remain unrecognized as such among the many that have not been stabilized.

Early rammed-earth structures include the Alhambra, Spain, built in the 14th century. (See photo 1c.) The conservation architect Pedro Salmerón, who has recently held a number of senior positions in connection with historic buildings and lectured in the history of architecture at the School of Technical Architecture, Granada University, gives details of materials used for building the Alhambra in *The Alhambra: Structure and Landscape* (2007).

The basic material, *alpañata*, is described in his book as 'intense red-coloured clay found in the natural Alhambra ground (Alhambra conglomerate). It is an important part of the earth-based walls of the Alhambra and gives them their characteristic red colour. Mixed with lime and sand, it is also used in the Alhambra footpaths.' Captions to photographs in his publication confirm that the local conglomerate soil – comprised of clay, gravel and sand – has a binding quality that makes for very stable construction and is used to make walls by adding materials such as lime and gypsum. The author shows photographs of the lime-stabilized and compacted Alhambra earth walls, and outlines the application of original lime render.

Salmerón states that 'Alhambra conglomerate can be mixed directly with 20% lime to form an "adobe" wall of an extremely rich quality whose secret is in the meticulous composition, using a dense, watered mixture in thin layers of about 10 cm (4″)' (Salmerón, 2007: 50). He confirms that the composition of the *alpañata* conglomerate is approximately 70% quartz, 20% clay and mica minerals, and 10% other components. The quartz aggregate is described as having 'a heterogeneous but balanced composition which includes gravel, medium-sized stones, and strong sand-like particles' (See photo 1c).

Records of the early influence of the Portuguese in South America confirm the tradition of earth stabilization with lime. The first national capital of Brazil, Salvador, was founded in 1549. Mário Mendonça de Oliveira from the Federal University of Bahia Faculty of Architecture, Salvador, submitted a paper in connection with lime-stabilized soil research to the New Mexico 1990 conference on the Conservation of Earthen Architecture. The paper refers to a number of early records including one from 'Luiz Dias, master of works responsible for planning and construction of the city of Salvador' in the 16th century. Mendonça de Oliveira states that these records confirm that lime was used as a stabilizer from the beginning of colonization, and that this method of construction was brought to the colony by the Portuguese. The paper also describes discoveries of and research into lime-stabilized adobes during restoration of a 19th-century house on Cow's Island, Bahia. Archive records confirmed the dating and laboratory analysis by the university and gave a $CaCO_3$ (calcium carbonate) content of 20.5%. ASTM particle-size grading of the soil used for the blocks indicated an average composition of 13% gravel, 9% coarse sand, 13% medium sand, 22% fine sand, 21% silt, and 21% clay.

Following archival research, practical field work for architectural conservation and laboratory analysis, Mendonça de Oliveira concluded that even though some early *pisé-de-terre* walls were built without lime, the process of using lime as a soil stabilizer for wall construction was brought to Brazil in the 16th century.

Evidence that the lime stabilization of compacted-earth building elements was common practice for the Portuguese is supported both by the Brazilian examples and by recent analysis of rammed-earth buildings in Spain and Portugal. Research completed in 2014 by Mileto and Vegas confirms that most of the compacted-earth walls in the Iberian peninsula contain 10%–20% lime, and that the highest proportions are found in Portugal.

2.3 Earth floors

In *The Development of English Building Construction* (1916), C.F. Innocent writes of improvements that took place when 'lime, a material with valuable setting properties, was mixed with floor material'. This is in the context of rammed earth and clay mixtures for solid floors. He also notes that floors containing lime seem to have been in use from the Norman period of the 9th and 10th centuries (that is, nearly a thousand years ago).

Jane Fawcett (1998) describes two mediaeval lime-stabilized earth floors. The first is in a town house in French Street, Southampton, where the original floor of lime-stabilized clay soil was replicated during repair and maintenance to make up losses in decayed areas. The second, at 12th-century Fountains Abbey, has been maintained and used after archaeological investigation. Research confirmed that the surface of the original was finished with a hard (well-compacted) mixture of lime and clay-based soil.

Research published in 1992 for the Aga Khan Cultural Services, Zanzibar (3D design-development-display OEG, 1992), in connection with repairs to the 19th-century Ithnasheri Dispensary, confirms the findings of 1989 research carried out for the Intermediate Technology Development Group (ITDG) and the Stone Town Conservation and Development Authority (STCDA). Lime-stabilized lateritic soil had been used for a wide range of the dispensary's building elements, including floor slabs. Sections taken through the floors of the entrance hall, internal rooms, and the courtyard showed they were of similar construction and all composed of red earth (lateritic soil), lime particles, and coral ragstone. (See photos 5a and 5b.)

Small pieces of coral ragstone, known as *kokoti*, were used as aggregate. The floor sub-base and hardcore include additional, larger, pieces of coral ragstone. The lower layers of the floor structure, of which there are normally at least three, are composed of lateritic soil, lime, and *kokoti*. The upper layers are of increasingly small, dense particles – although some of the finishes may not now be original. Lime-stabilized lateritic soil appears in all layers except for the compacted hardcore. The finest aggregate, and probably pozzolan, are likely to have been incorporated in the original surface finish.

The lime stabilization of earth floors appears to have been practised worldwide for centuries. Eurwyn William, in his book *The Welsh Cottage* (2010), refers to John Evans' 1798 description of an earth floor for a cottage built in Merioneth, Wales: 'The floor was the native soil, rendered very hard and uneven from long and unequal pressure.' He confirms that a typical 'recipe' for a mud floor was four or five barrows of earth, one and a half buckets of lime and one barrow of cow dung. A further example of this type of floor is given by Ashurst and Ashurst (1988), who quote a traditional specification as 'one-third lime, one-third well-sifted coal ashes and one-third loamy clay and horse dung made from grasses. Mix the materials in a dry place, let it stand for ten days then mix again, adding a small quantity of water. Let it stand once more for three to four days, mix once more, then the mix is ready to be laid.'

This material needs to be well and evenly compacted for best results and durability. Ashurst and Ashurst also advise that a clay-rich earth floor may be stabilized by raking in 10%–15% hydrated lime, watering, and compacting.

2.4 Clay lump, earth bricks and blocks

There is a long history of using various types of earth block, including clay lump, slop-moulded earth bricks, and sun-dried compacted blocks, for buildings worldwide. Although un-stabilized earth bricks and blocks have often stood the test of time, flood disasters have demonstrated the need for stabilization, particularly in areas subject to long-term saturation. Archive records about stabilizing earth blocks are scarce, although the example of lime-stabilized adobe blocks for the 16th-century house in Bahia described above indicates that it was common practice in some communities.

In 1726, Nathan Bailey and John Worlidge, writing in *A Dictionary of Husbandry, Gardening, Trade, Commerce and All Sorts of Country Affairs: Volume I*, stated: 'Take loam or a brick-earth, and mixing therewith some good lime, temper them very high till they become tough, smooth and "glewy"; let the wall of your house be one brick or one and a half thick, and your unburnt bricks being laid in this well-tempered mortar, they will cement and become one hard and solid body.'

In the introduction to his *Conservation of Clay and Chalk Buildings* (1992), Gordon Pearson confirms that lime putty can act as a stabilizer. He claims that when it is added to a chalk wall, the chemical reaction is considerable and leads to early strength.

A knowledge of ways in which earth bricks and blocks may be stabilized is useful for conservation as well as being important for designing mixes for durable and low-cost building in flood-prone areas. Field-test methods for lime-stabilized earth blocks are given in Section 5.8. Small-scale block production is described in Section 6.3.

2.5 Wattle and daub

There is a long history of constructing wall panels by applying modified clay-rich earth (daub) to interwoven sticks or laths (wattle). Various materials added to the mix improve the daub's workability, adhesion, and weathering qualities. Writing about 16th-century daub in Sheffield, Innocent (1916) states that 'materials used for daub and plaster were closely similar to those used for mortar', which he describes as a mix of clay, mud, lime, and cow dung. He goes on to say: 'The old builders (1796), knowing nothing of the chemistry of mortars, and making little distinction between lime and other materials, made improvements by trial, error, and experience.' Thus, although lime was sometimes added to daub mixes, there was no certainty of stabilization. An understanding of the chemistry, mix proportions, and best practice is required for predictable results, and the following chapters set out our contribution to this.

Kenneth Reid (1989), writing in a technical pamphlet for the Society for the Protection of Ancient Buildings (SPAB), gives some examples of daub mixes once used in the UK:

- *8 parts stiff sandy clay soil, 1 part lime, 1 part cow dung, and 1 part straw*
- *4 parts sandy clay soil, 1 part cow dung, 1 part lime and chopped hair*
- *3 parts reused old daub, 1 part lime putty, and 1 lb (pound by weight) of hair per cubic foot.*

(A cubic foot is 300 mm × 300 mm × 300 mm, so this is 16 kg/m^3.)

Recent examples of lime-stabilized daub include that used for rural housing in Zanzibar in the 1990s. The local clay-rich lateritic soil is stabilized with quicklime produced by burning coral ragstone from the island's quarries. After mixing and tempering, the mix is applied to a double grid worked from coppiced tree stems, and used as a daub with ragstone aggregate, to fill and consolidate the walling.

Repairs to a 16th-century timber-framed hall house in Stoneleigh, Warwickshire, have been described by the author (Holmes, 2002). The timber frame was in-filled with wattle and daub panels. Some of these required repair and a few needed to be completely replaced due to defective thatch and previous inappropriate repairs. The original daub was analysed and found to contain local clay, a range of fibres including chopped straw and hair, and a small percentage of lime. A matching lime-stabilized daub mix, which included 5% finely powdered quicklime, was prepared and applied immediately after mixing in the quicklime to make good the missing and defective areas. The repaired panels were finished with limewash on lime render, and continue to remain sound in a moderately exposed location, in UK weather conditions after 17 years. (See photos 3a and 3b.)

2.6 Earth mortars

The use of earth mortar was a common practice in the British Isles throughout the mediaeval period. Its application continued into later centuries but here, too, improvements were made by trial and error. Experiments with the addition of other materials only later became more scientific. At first, other materials were chosen mostly from those that were both close at hand and which had previously proved beneficial. One of these materials was lime. Over centuries, at different times and with different rates of change in various parts of the country, earth mortars and daubs were improved. These improvements were in parallel to the development of lime–sand mortars that were preferred and used primarily for higher-status buildings and areas of severe exposure. The early use of lime-stabilized soil for mortar should therefore be expected where both clay-rich soil and limestone are readily available local resources.

The Plymouth Building Accounts from the 16th and 17th centuries include a description of the mortar used for Plymouth's Guildhall. This is summarized by E. Welch of the Devon and Cornwall Record Society, as:

> *For the new, later Guildhall, however, lime, sand and earth is carried to site, suggesting earth–lime mortars were improved before use by the addition of sand. The evidence of the Shambles and later Guildhall accounts, is that the mortars were both earth–lime and lime, the former probably forming the bedding mortar and possibly the base coat plasters, although when plasterers are being paid, lime, lime–ashes, sand and hair are also listed.* (Welch, 1967)

There are many examples of lime-stabilized soil mortars throughout the UK. One is at the National Trust's Holnicote Estate in Somerset, which includes several villages and farms. The area contains clay-rich earth and outcrops of blue lias limestone (beds of naturally occurring hydraulic limestone), and there is evidence that limestone was also shipped across the Severn Estuary from Wales to lime kilns on the Somerset and Devon coasts. The advantages of using lime, as either quicklime or putty, with the local soil were discovered, and there are examples of its use in stabilized-soil mortars and plasters where it remains sound in farm buildings and 18th- and 19th-century houses on the estate that are still occupied today (See photo 6).

There are also numerous examples of earth–lime mortars in both high-status and smaller domestic buildings in Malton, North Yorkshire. These buildings date mainly from the 16th and 17th centuries, and currently remain occupied and in regular use. It appears to have been common practice in the area at that time to construct stone buildings using earth or earth–lime mortars. Some earth–lime bedding mortars were finished with rich lime-mortar pointing. These include York House and the Talbot Hotel in Malton, which are some of the higher-status buildings in the area (Copsey, 2016).

The compressive strength of lime-stabilized soil mortar can match or exceed that of a non-hydraulic lime–sand mortar. Stabilized-soil bedding mortar and wall cavity fill are better protected from the elements than pointing mortar, and may last for centuries. They are an important part of a building's fabric and history. Hopefully, greater recognition of this material will lead to less of it being removed in error and more being conserved.

2.7 Internal wall and ceiling plaster

In buildings where lime-stabilized soil has been used for one or more elements, it is likely that wall and ceiling plaster will be among those elements. Some historic buildings in which lime-stabilized soil has been found and conserved in wall and ceiling plaster are:

- Motslow Hill, Warwickshire (16th century)
- Fort Jesus, Mombasa, Kenya (16th century)
- Listed Manor House, Devon (17th century)
- York House, Malton, Yorkshire (17th century)
- Kot Diji Fort, Khairpur, Pakistan (18th century)
- Holnicote Estate, Somerset (18th and 19th century)
- Ithnasheri Dispensary, Zanzibar (19th century)
- Houses at Vilamar, Portugal (20th century) (Fernandes, 2008).

Internal surfaces are usually finished with a thin (0.5–2 mm, $\frac{1}{32}''$–$\frac{1}{16}''$) lime-rich skim coat or limewash, and sometimes with a thicker coat of lime render. Sharp well-graded sand and fibres, a hot or warm climate, good compaction and effective curing all contribute to durability. It appears that lime used in conjunction with soil in this way for plaster has been practised worldwide from at least the time of the Alhambra in the 14th century. Whilst the method was not fully understood or universally used, the material was applied where local conditions were favourable – the presence of clay-content soil and lime being the most important of these. (See photo 2.)

Traditional mixes used for internal wall finishes in parts of Wales and recorded by Eurwyn William (2010) include equal parts loam and lime (for best quality) and two parts loam to one part lime (for a lower-cost finish). Citations for loam in the *Oxford English Dictionary* include 'a composition of moistened clay and sand with chopped straw, etc. used in making bricks, casting moulds, plastering walls, grafting, etc. (1395)' and 'a rich soil composed chiefly of clay and sand with an admixture of decomposed vegetable matter (1664)'. In *An Encyclopaedia of Architecture* (Gwilt, 1894), it is described as 'a soil in which clay prevails. It is called heavy or light as the clay may be more or less abundant.' Writing in 1726, Richard Neve describes loam as 'a sort of reddish earth, (well known), used in buildings (when tempered with mud gelly, straw and water) for plastering of walls in ordinary houses'.

Obtaining lasting results with lime-stabilized soil plaster requires a good understanding of the materials and having the skill to apply it. Well applied and tended, it can survive for centuries, as many existing buildings demonstrate, including those described above. In addition to sound craft skills, the selection of the right materials and their quality, preparation and tending are important when aiming for the best results and durability. The quality and types of lime, soil, clay, sand and fibres – possibly also pozzolan – need to be known and carefully selected, as do the mix proportions. Field tests for these are set out in the following chapters together with recommendations for preparation, application, and aftercare. (See photos 4a, 4b, 5a and 5b.)

2.8 External renders and screeds

External renders are exposed to all weather conditions as well as impact and abrasion, so need to be strong, impact resistant and durable. At the same time, it is important that they adhere to and are compatible with the surface to which they are applied. If they are to be used with earth structures, they therefore also need to be relatively soft, pliable, and vapour permeable (whilst remaining wet-weather resistant). Some of the most complete and extensive remains of lime-stabilized soil renders are on early fortifications. Examples include:

- Alhambra Palace, Grenada, Spain (13th and 14th centuries)
- Makli Necropolis, Sindh, Pakistan (14th century)
- Fort Jesus, Mombasa, Kenya (16th century)
- Kathmandu Temple, Nepal (18th century)
- Zanzibar Fort and other buildings in Zanzibar town, including the Ithnasheri Dispensary (17th to 19th century)
- Kot Diji Fort, Khairpur, Sindh, Pakistan (18th century).

These buildings have escaped much intervention, including the process of re-rendering with the stronger and more recently developed Portland cement-based materials, due to a combination of factors: the forts have been redundant and mainly disused for a considerable time, access for maintenance is difficult and costly, and some are well-protected from severe exposure to the elements (by, for example, well-maintained roof overhangs, as in Nepal), which helps to improve the longevity of a render. (See photo 16c.)

Original internal plaster and wall paintings from the 16th century also survive at Fort Jesus. Whether some or all of the lime-stabilized external earth renders are original is not certain. It is clear from archaeological evidence, however, that they have remained in place for many years, and, in some cases, for centuries. (See photo 1b.)

2.9 Surface finishes

The principal surface finishes considered here are for walls, floors, and roofs. Used in conjunction with resilient external renders, all other external surface finishes need to be durable. Whilst many materials have traditionally been applied to renders and other surfaces to improve durability, it is unlikely that there is a more compatible finish for a lime-stabilized soil surface than a carefully prepared and applied limewash. Limewash is suitable for well-finished, un-rendered lime-stabilized soil building elements such as blockwork, cob, and *pisé-de-terre*, for rendered and plastered walls and ceilings and for daub. Examples of the durability of well-applied limewash as an external finish may be seen on the buildings described above, as well as on many other historic and occupied buildings today. (See photo 3a.)

2.9.1 Internal walls and ceilings

Because internal wall and ceiling surfaces are not exposed to the elements, their finishes do not require the same level of durability as external walls. In addition to limewash, Gordon Pearson (1992) refers to casein, tallow, and linseed oil being used in internal decoration, and states that chalk slurry mixed with limewater was a common finish in areas where chalk was readily available, particularly in the Wessex Downs. It was typically made with crushed chalk – washed, dried, ground fine, and bound with diluted size. Finely powdered animal glue or keratin dissolved in hot water is the basis of distemper, which was a widely used internal finish. It is useful for surfaces that remain dry and are exposed to little wear, typically ceilings. Whilst not as durable as limewash, durability can be improved by the addition of diluted soap, producing a more lasting 'hard distemper'.

For colouring limewash and clay paints, natural (earth) pigments are likely to be a more compatible and safer choice for historic building fabric and lime-stabilized soil substrates than proprietary products based on impermeable, chemical formulations, now produced on an industrial scale worldwide. Many of these formulated surface treatments are not suitable for applying to a soft and permeable background, largely due to the differences in permeability.

2.9.2 External walls

Accounts of clay soils or loam used for plastering walls up to the 19th century suggest that, although the chemistry may not have been fully understood, the benefits of combining a limewash or lime render finish with a clay-soil background were well known. Lime stabilization of the soil at its interface with the surface coating would provide both a good long-term bond and improve weather resistance. Durability may have been improved by trial and error, the type and proportion of clay in the soil chosen, and the use of a natural hydraulic lime or lime-stabilized soil. This was often governed by local availability but considerable adjustment of materials and mixes would have been possible. (See section 6.12 on testing finishes.)

2.9.3 Floor finishes

Writing in 1726, Neve describes floors as being 'of several sorts; some are of earth, some of brick, some of stone, and some of wood [...] earthen floors are commonly made of loam and sometimes (for floors to make malt upon), of lime, and brook-sand, and gun-dust, or anvil-dust from the forge' (Neve, 1969).

The choice of floor finish is governed by a number of factors, including requirements for durability, cost, and appearance. Price and availability influence the choice of materials, whilst the quality of craft skills dictates the standard of finish. Earth floors can be level, hardwearing, and well-finished provided they are laid with care and skill. Records show that the finished surface may be hardened in several ways. The inclusion of fine and hard aggregate, without reducing the clay proportion of the mix, combined with thorough and forceful compaction, improves impact resistance. Several 18th-century authors record the benefits of tempering fine clay with ox blood to achieve a strong, well-bound surface. There are many other methods recorded, an example of which is from Eurwyn William (2010), who recounts: 'The inhabitants of Glamorgan refined the art of making floors by adding lime or crushed seashells in the coastal areas. They perfected mortar floors capable of lasting for decades, if not centuries, which presented an entirely smooth surface when newly made. [...] The floor could be washed with sooty water to harden it and make it shine.'

Beeswax and turpentine are traditional and long-lasting surface finishes suitable for unpainted timber, as are various natural oils, typically linseed oil, for improving the water-shedding and weathering properties of external surfaces. More locally available oils include mustard oil in Nepal. Recommendations given by the authors of *Building with Cob* (Weismann and Bryce, 2006) for finishing earthen floors with beeswax and turpentine are in accordance with this long-established practice.

They advise that following thorough preparation and compaction of the sub-floor layers, the final surface should be hard-polished by damping and scouring in a similar way to that used for a plaster (lime–ash) floor. When it is dry, it is ready for sealing with linseed (or other) oil and thinners such as turpentine or citrus oils, and possibly finished with beeswax (see section 6.11.1).

CHAPTER 3
Building limes and hydraulic set

3.1 The lime cycle

The lime cycle explains one of the many environmental benefits of using lime outlined in the first chapter. When lime is used in buildings it eventually reverts to calcium carbonate, which chemically is the same as the material from which it was originally prepared. Most of the carbon dioxide gas driven off from limestone during lime burning is eventually replaced (re-absorbed) from the atmosphere, in a process known as carbonation. The full sweep of the cycle is the conversion from calcium carbonate ($CaCO_3$) to calcium oxide (CaO), followed by combination with water (H_2O) to form calcium hydroxide ($Ca(OH)_2$). Finally, through carbonation, water is lost and carbon dioxide regained, to re-form calcium carbonate again (see Figure 3.1 and photo 10).

The lime cycle is therefore a journey of transformation, during which calcium carbonate in the form of hard stone (limestone) is turned into a mouldable form (powder or putty), to be mixed with soil or sand (or both) to make building elements, and then reverts back into calcium carbonate as part of the building. It sounds like the magic of alchemy, but this is not a cycle of turning one material into another; it's a cycle of turning the raw limestone into a mouldable and more useful form to bind and protect our buildings. This process of converting a hard stone or seashell into durable, protective mortars and renders is part of the magic of lime, and has been practised and used as such for thousands of years. Most limestone is a sedimentary or metamorphic rock produced over geological time, formed from living organisms. Lime is life!

The seashell is a clue as to the origins of lime. Most limestone, one of the most abundant minerals on the planet, is effectively sediment made up primarily of the calcium-rich skeletons and shells of sea creatures. These sediments were laid down millennia ago in layers of what eventually, through various geological processes, became calcium-rich stone. It is often possible to see some of the fossils of these ancient sea creatures in the limestone (Appendix 3).

Calcium-rich limestone is known as pure lime, as the sediments are made up almost entirely of calcium carbonate with very few impurities. Pure lime is also known as non-hydraulic lime, which is a reference to its inability to set or harden under water. This is in contrast to the majority of hydraulic limes, which will set under water.

Many limestone sources will produce a non-hydraulic lime. This can be of very high quality and has many uses other than for building, including

26 BUILDING WITH LIME-STABILIZED SOIL

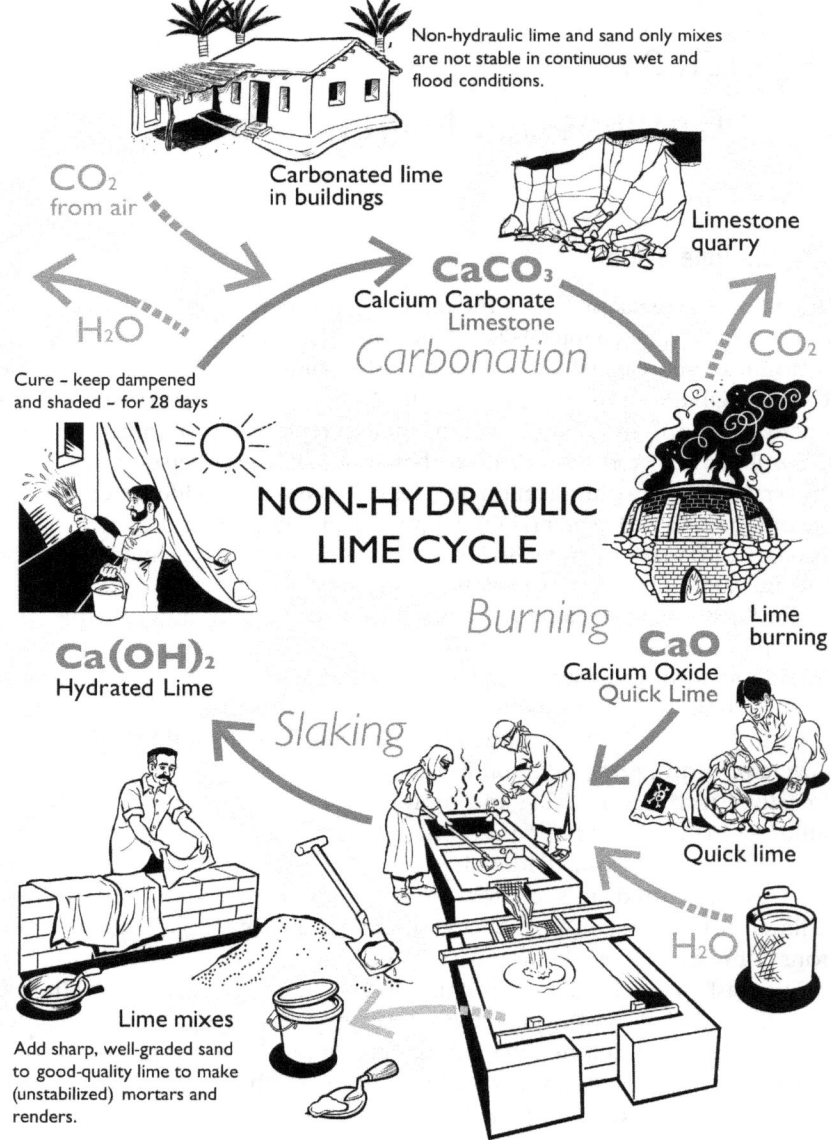

Figure 3.1 The non-hydraulic lime cycle.

purification of water, medical treatment, agriculture, food processing – such as sugar refining – and industrial uses.

A hydraulic set is needed for resilience against wet conditions, particularly standing water after floods, and for subfluvial structures such as dams and canals. Building elements exposed to water may need to stay stable in wet conditions for months, and sometimes permanently.

The first stage of field testing examines methods of preparing and testing individual materials that will contribute to a hydraulic set with non-hydraulic lime, and are appropriate for a range of building elements.

3.2 Hydraulic set

In the context of non-hydraulic lime as the base material, the term 'hydraulic set' refers to the action that results from combining non-hydraulic lime with minerals, in the form of either 'active clays' or those of a pozzolan, which may be artificial or natural, enabling the resulting material to become insoluble, resist damp conditions and remain set (or stable) under water. The term 'active clay' was proposed by Vicat (1837) as a reference to the type of clay that will react with lime to provide a hydraulic set and is a simple term convenient to use in this context.

Active clay mixed with lime and water is the principal way to produce the set required for stabilization. The primary minerals present in active clays are usually alumina and silica and sometimes also iron oxide. They are found in clayey soil and some limestones.

Lime stabilization of soil is largely based on the chemical reactions between the lime and the soil's clay minerals. Clay in the soil is modified because of the action the lime has on, and between, the clay molecules. These molecules are less than 0.002 mm (2 microns) across and are produced by the chemical weathering and leaching of minerals from rock formations. They have a unique chemical composition and specific physical properties.

The fine grains in clay are formed of irregular crystals, of which many are hexagonal. These consist of many thin sheets, or layers, which can contain from tens to hundreds of individual atoms. Most are molecules of silica or alumina, surrounded by oxygen atoms (Houben and Guillaud, 1989).

Characteristic chemical reactions occur in the first stage of the stabilization process. Since some of these are directly related to the laminar structure of each clay, and to the spaces between their layers, the reactivity of clays varies depending on the type and the spaces between layers. Crystals within the clay molecules may be negatively or positively charged, so the cohesive forces between them are largely electrostatic. In the presence of water, where cations originating from the introduced lime saturate most of the soil's minerals (including the potassium and sodium ions), the properties that depend on the electrostatic charge and limits of consistence are suddenly modified. The interchange of cations increases bonding between the clay particles, which flocculate to form coarser agglomerates. The initial fast reactions include absorption of calcium hydroxide molecules and ion crowding.

Secondly, absorption of calcium hydroxide produces calcium silicates, calcium aluminates and other compounds of hydrated calcium. These reactions may happen quickly and can produce a calcium silicate hydrate gel and other cementitious binding materials, which will set over a period of time depending on local conditions. Water in connected pores between grains of the compacted matrix assists reactions and an increase in the solid components of the gel. Aluminate reactions are slower: calcium aluminate and calcium alumina silicate crystalline phases occur later in the setting process, which may extend up to three years, or longer.

As more solids are produced, the bond strength of the lime-stabilized soil increases and new compounds grow in the pore space. This results in a progressive reduction in permeability (Bectham, 2010) and an increase in compressive strength.

Carbonation also occurs over time, leading to a further slow increase in the bond strength of the cementitious materials in the presence of moisture. However, it is temperature-dependent: there may be little to no carbonation below 5°C, but as the temperature rises to 10°C and above, the rate accelerates.

To summarize, the three principal mechanisms of bonding and set during the stabilization of soil with lime are therefore:

- *fast reactions* – cation exchange resulting in modification of the clay's structure;
- *slower reactions* – development of calcium silicate, calcium aluminate hydrate gel and other cementitious compounds increasing bond strength; and
- *medium- and long-term reactions* – carbonation of calcium hydroxide developing into calcium carbonate, leading to hardening, long-term calcite growth and increased compressive strength. (See 'Stabilization' in the glossary.)

These mechanisms occur simultaneously, but at different rates and for different lengths of time. The compressive strength of lime mortar, for example, may continue to develop over time, and after two years it can be twice that at 28 days. Time, temperature, and moisture have a direct effect on these reactions: the strength of lime-stabilized soil increases with extended time, at higher temperatures, and in the presence of water. Final strength is also greatly improved by initial compaction. Using quicklime, as opposed to lime putty or dry hydrate, accelerates the process. Uniform dispersal of the lime, adequate moisture, and full slaking of the mix are also important for assisting these reactions.

3.3 Natural hydraulic lime

Natural hydraulic limes are made by burning limestone in which active clay was laid down at the same time as the calcium-rich remains of sea creatures. These eventually combined to form a less pure limestone than those used to create non-hydraulic (pure) lime, because of the additional minerals in the sediment. It is some of these minerals or active clays that are essential for creating a set under water; the clays and the lime combine chemically when this type of limestone is burnt, to produce a natural hydraulic lime.

BUILDING LIMES AND HYDRAULIC SET   29

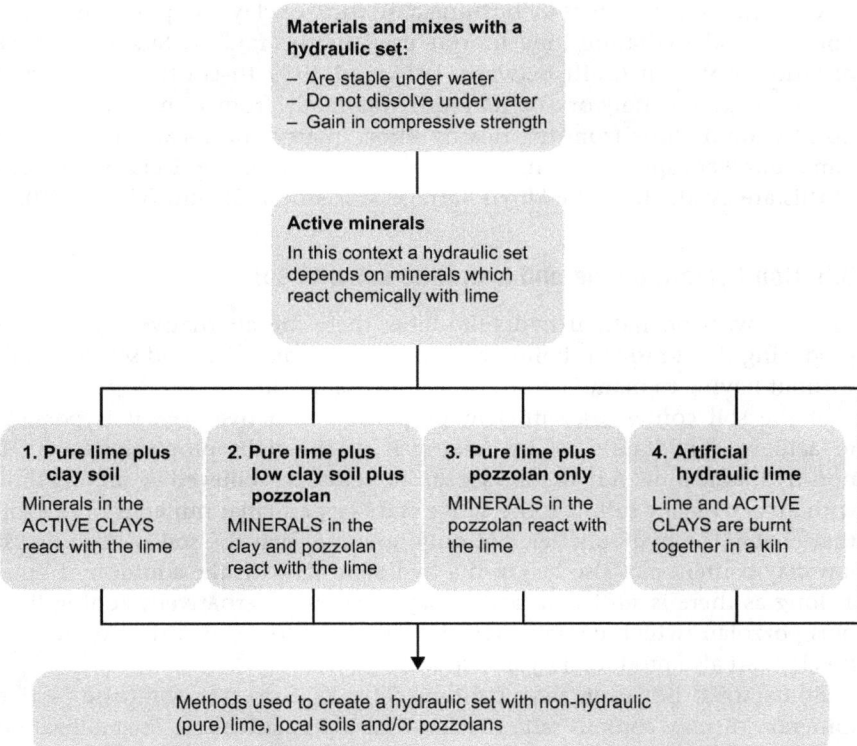

Figure 3.2 Four ways to create a hydraulic set with non-hydraulic lime.
(Hydraulic set can also be achieved with a natural hydraulic lime, in which active clays had formed part of the limestone.)

Different strata of some limestones and different limestone formations will contain varying amounts of active clay, which will determine the degree of hydraulicity of each. There are national classifications for natural hydraulic limes which cover a range of hydraulic set – from weak to very strong. The main three classifications of natural hydraulic lime were traditionally known as feebly, moderately, and eminently hydraulic limes. The stronger eminently hydraulic limes, sometimes with pozzolan added, have been used in mortar for water mills, river embankments and lighthouse walls that are permanently under water – sometimes in forceful, running water (see photo 7).

3.4 Artificial hydraulic lime

As in Figure 3.2, where the limestone is almost pure and cannot produce a natural hydraulic lime, a hydraulic set may be achieved by using the non-hydraulic lime and combining it with an active clay from another source. This is the principal method of obtaining a set with artificial cementitious binders.

One way this is done is by burning a mixture of clay and pure limestone. The artificial hydraulic lime mortar used in the famous Sukkur Barrage of Sindh, Pakistan (built between 1926 and 1939 to control the world's largest irrigation network of its kind) was made from a mixture of pure non-hydraulic lime from the nearby Rhori quarry and local clay burnt in kilns that were specially constructed on the building site. Detailed records of this are available at the Lloyd Barrage Museum in Sukkur (Hill, 1920).

3.5 Non-hydraulic lime and hydraulic set with soil

In areas with no natural hydraulic lime, there are alternative methods of producing flood-resilient buildings with low-cost materials and selected soils without having to manufacture artificial hydraulic lime.

If the soil contains a sufficient proportion of active clay, it is possible to achieve a hydraulic set by mixing it in the right proportions with a non-hydraulic lime. A hydraulic set can therefore be achieved by mixing lime with an active-clay soil. It is the active clays or particular minerals in the soil that enable the hydraulic set, not only how clay-rich the soil is. A soil with low clay content can also be given a hydraulic set with the addition of lime, as long as there is sufficient 'active clay' present. Alternatively, adding lime plus pozzolan (which contains reactive minerals) to a soil with very little or no clay can also produce a hydraulic set.

Some soils, however, may contain clays that do not comprise 'active minerals' or may contain salts that prevent a permanent set or stabilization with lime. This should be determined at Stage 2 field testing (see Section 3.7 for more detail and Chapter 5 for field tests). If a soil cannot be stabilized, then it may be possible to locate an alternative local soil source, as soils often vary over short distances and depths.

Local soils can be investigated and tested to determine the quantity of non-hydraulic lime that will be needed to formulate mixes that will remain stable under water; to establish appropriate particle-size grading for the building element concerned; and to assess whether it will be necessary to add a pozzolan.

All mixes for the building elements and components intended to remain stable under water need to be tested to determine the variation produced by local soil types and pozzolans. Unless full laboratory analysis is possible, it is advisable to carry out all three field-testing stages.

3.6 Pozzolans

Since they lack the active minerals found in clayey soil, low-clay soils, sandy soils, sand and other aggregates will need a pozzolan (sometimes described as pozzolana) to be added for stabilization when they are mixed with a non-hydraulic lime. This resultant pozzolanic reaction with lime is another way of creating a hydraulic set.

BUILDING LIMES AND HYDRAULIC SET   31

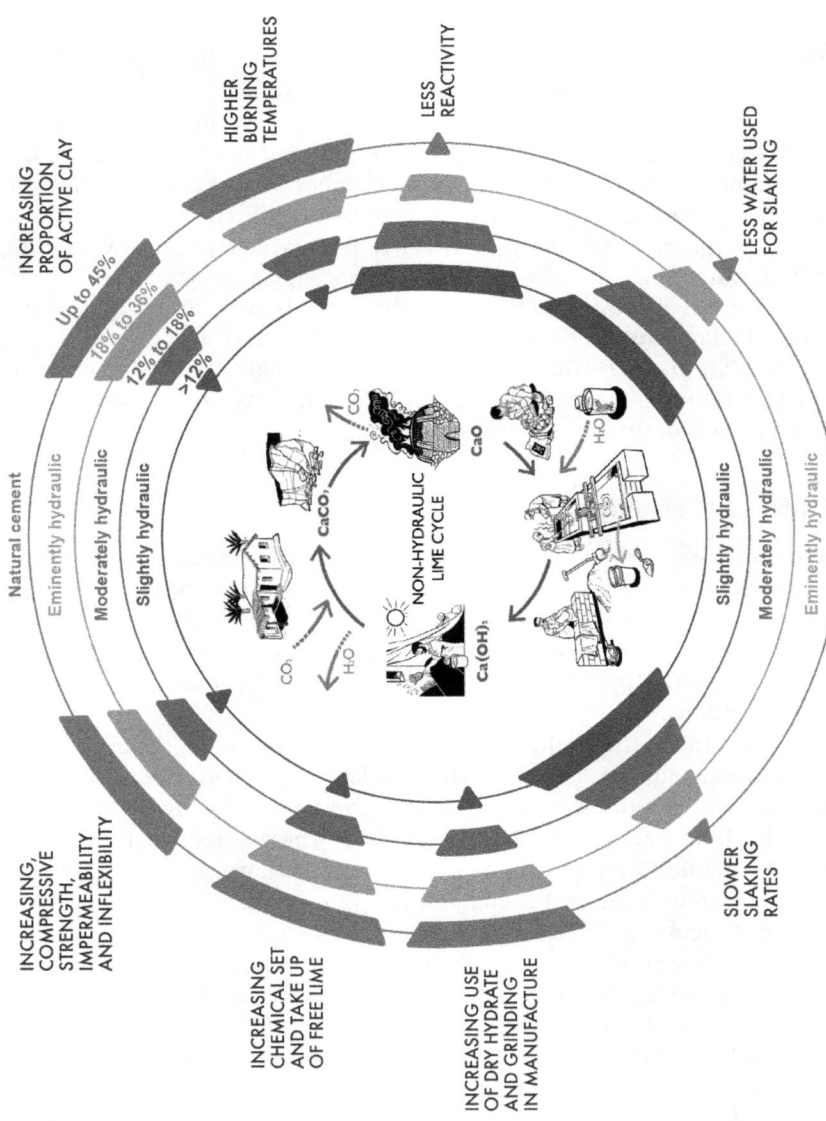

Figure 3.3 Variations to the lime cycle produced by hydraulic limes. (See photo 7 for this diagram in colour.)

The term pozzolan is derived from Pozzuoli, the name of the town in Italy which was an important source of the sand and volcanic ash used by the Romans to create hydraulic mortars over 2,000 years ago. The Romans were highly skilled water engineers, and many of the baths, aqueducts, and bridges they built using mortars of non-hydraulic lime and pozzolan remain standing. Some are still in use today.

There is a variety of pozzolanic materials, including finely sieved burnt-brick dust, rice-husk ash, pulverized fuel ash (PFA), other industrial wastes and natural volcanic deposits. The exact ways in which many of these materials react with lime to produce a chemical set are variable, but they can be predicted following detailed laboratory analysis. However, field testing can give a sufficient indication of their reactivity for practical purposes in stabilizing low- or non-clay-content soils with pure, non-hydraulic lime.

To be at their most effective, pozzolans need to be finely powdered. Soft-burnt clays (usually high in silica and alumina) and rice-husk ash particles (high in silica) will normally require thorough crushing or grinding and sieving, although larger particles may be incorporated as aggregate, which would have less or little pozzolanic effect.

3.7 Choice of materials

The reason and principles for the selection and testing of materials described here are for the purpose of producing reliable and stable building elements.

A careful choice of materials is critical, particularly where options are limited and where a hydraulic set is essential for the construction of buildings in wet conditions.

In many parts of the world and often in flood-prone areas, the application of appropriate, low-cost, and locally available materials is the best and sometimes only option, so field tests and methods for assessing the suitability of soils for lime stabilization are important. In these cases, the preparation of non-hydraulic limes and their use with either clay soil or pozzolans to produce a hydraulic set are fundamental requirements.

Details of field tests for a range of materials, including lime, clay soil, and pozzolan, are given in Chapter 4. Chapter 5 describes site-based preparation of lime-stabilized soil test mixes, and testing of lime-stabilized soil building elements is covered in Chapter 6.

The field-test methods described are an initial guide and an immediate way forward for the lime stabilization of soil needed for both urgent repair and rebuilding programmes. Due to the varied nature of soils, however, it is best if the field tests are followed by or carried out in conjunction with laboratory analysis to establish the precise chemistry and other characteristics of the soils, particularly for large-scale and long-term construction projects. Until laboratory test results are available, unsuitable soils may be identified by visual inspection of the raw material, the use of reagents for limited chemical analysis, long-term field testing for durability and strength, and measuring the expansion of test samples.

3.8 Three-stage field testing

Due to the extreme variation of soil types, it is recommended that an initial field-test programme – prioritizing lime stabilization – is undertaken to identify, prepare, and test materials, mixes, and building components. This process has up to four stages, with the first three being in the field:

1. The first stage is to test individual primary materials for suitability: building limes for reactivity, soils for particle size and clay content, pozzolans for reactivity, and fibres for appropriate size and strength. This is covered in Chapter 4.
2. The second stage is to test well-prepared mixes of materials that the first stage of testing has shown are of a satisfactory standard, primarily to ensure that they will remain stable under water. Fully cured trial mixes should be tested to determine the most suitable and effective mix for each building component, particularly the optimum lime proportion for soil stabilization, before any are used in the main work. This is covered in Chapter 5.
3. The third stage is to test completed samples, incorporating all the materials and modifications for the proposed building components for the main production run. The sample components will be based on and modify the successful mixes during site trials, for each specific building element and application before their final manufacture. Testing should be continued at intervals throughout the main work to check for quality and consistency. This is covered in Chapter 6.
4. A fourth stage of investigation is the laboratory testing of final mixes. This should be carried out before the main work if practical, and if there is sufficient time and funding. This would involve soil particle size, chemical analysis, and measurements of unconfined wet and dry compressive strength (UCS) to test long-term durability. This is discussed with explanations and references to current civil-engineering practice in Chapters 7 and 8.

The three-stage field-testing programme may seem a long process, but once successful materials and their mixes have been identified, careful replication of the same mixes with similar materials from a particular region may be repeated indefinitely. The processes need to be well recorded and demonstrate consistent, reliable results. Ideally, field testing, particularly that of soils, would be immediately followed by detailed laboratory testing to ensure that only suitable or suitably modified soils are used. As outlined above, the variation in soil types is such that some may be unsuitable because, for example, they contain salts or agricultural run-off that may cause defects or decomposition. Such soils must be identified for modification or substitution prior to commencement of the main work.

Whilst the remaining chapters provide full details, the following pages and Figure 3.4 summarize the three field-test stages.

3.9 Three-stage field-testing summary

3.9.1 Stage 1: Investigate, test, and select individual materials

Test each material for quality and suitability:

- *Lime* – test for reactivity, fineness, and density.
- *Soil* – test for clay content and particle size.
- *Pozzolans* – test for particle size and reactivity.
- *Fibres* – test for size, strength, and record of durability.
- *Salts* – test soil and water for salts and sulphate content.

3.9.2 Stage 2: Prepare, test, and select mixes for stabilization

Using samples of the best quality materials that have passed the Stage 1 field tests, make three trial samples of each of three different trial mixes per building component with varying lime proportions (see Section 5.4). Fully cure the trial-mix samples for 28 days then immerse them under water to test for stability, and test for additional qualities appropriate for their use. For example:

- *Foundations* – compressive strength, aggregate size and proportion, and stability under water.
- *Wall blocks* – as for foundations plus tensile strength.
- *Mortar* – as for foundations plus workability.
- *Render* – crack free, stability under water plus permeability.
- *Plaster* – as for render.
- *Floor screed* – as for foundations, plus crack-free and impact resistance.
- *Roof screed* – stability under water and permeability.

3.9.3 Stage 3: Field testing of building components

Without changing the lime to soil proportions that gave successful stabilization in Stage 2, modify mixes as appropriate by the addition of materials such as sand, fibres, cow dung, and pozzolans. Test:

- *render and plaster panels* for workability, bond, crack resistance, and finish;
- *floor-screed panels* for crack-free robustness of finish and permeability;
- *roof-screed panels* for crack-free flexible finishes and permeability;
- *aggregates* for particle size, appropriate to the building component;
- *block, foundation, floor and wall mixes* for compaction;
- *limewash* for consistency, quality, adhesion and colour; and
- *all final building elements and components* for stability under water.

BUILDING LIMES AND HYDRAULIC SET   35

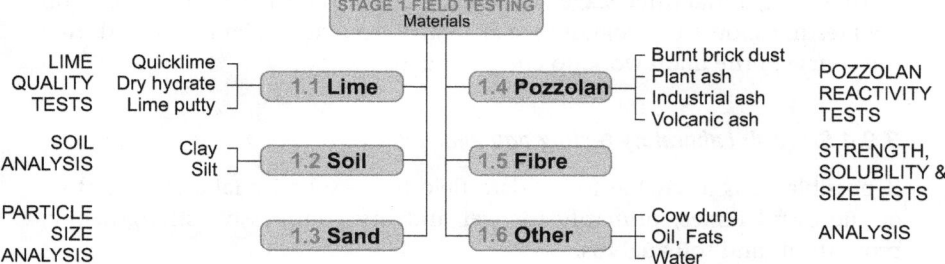

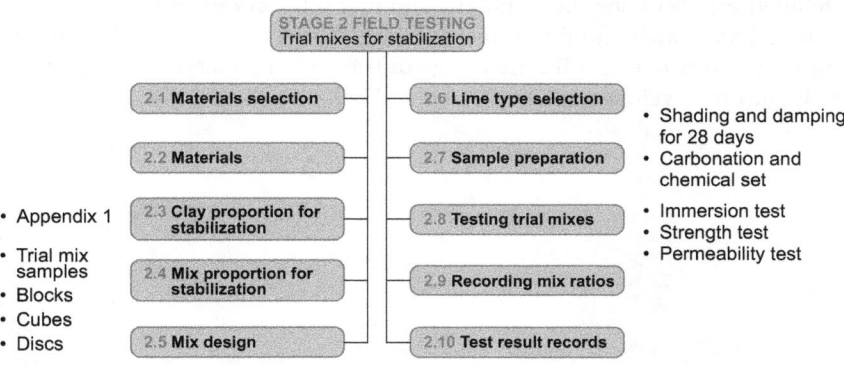

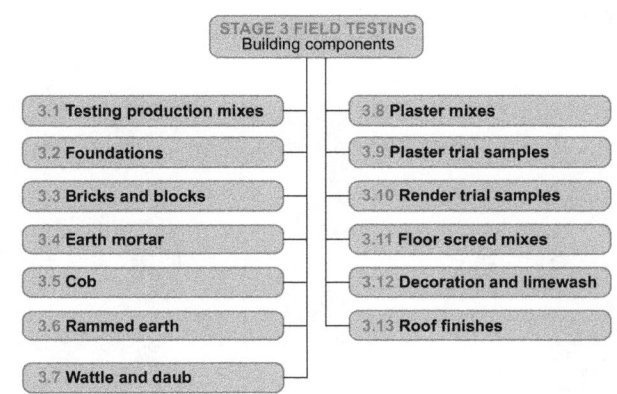

Figure 3.4 Flowchart of three-stage field-testing sequence.

36 BUILDING WITH LIME-STABILIZED SOIL

Following initial three-stage field testing prior to construction, longer-term field testing should be programmed and carried out at regular intervals during the course of the main work, to check for consistency.

3.9.4 Stage 4: Laboratory testing and analysis

If possible, it is advisable to validate field-test results by laboratory testing of successful mixes, primarily for wet and dry compressive strength, mix proportions and soil analysis.

The scope of this book does not extend to detailed descriptions of laboratory testing, although it is recommended that laboratory tests are undertaken on a selection of successfully field-tested mix samples (see photo 19). Detailed results of chemical analysis and mix proportions in accordance with national standards should be recorded for reference, and could be used to compile a database of soil mineralogy in the region concerned, which would assist future development.

CHAPTER 4
Field testing: Stage 1 – Materials

4.1 Building limes

4.1.1 General site preparation

Mixing yard. The mixing yard should be established near the lime putty slaking and settlement tanks and a ready source of water. It should be close to the construction site or block-making area, adjoining easy road access. Keep a clean, flat, secure area for all earth and lime mixing and tool storage. Ideally, this should be a hardstanding with a surface of stone or brick or other hard material, in order to avoid topsoil being inadvertently dug up and included in the mix. It is a working area, so protection from rain and sun is advisable: a covered area is recommended.

Block-making area. Establish a block-making and curing area alongside the mixing yard. It is best if the space for stacking and curing the lime-stabilized soil blocks immediately after production is long, flat, and shaded. In the tropics, prepare materials and coverings to provide essential shade well in advance (see Section 6.3).

4.1.2 General field equipment checklist for small-scale lime-stabilized work

Basic tools and equipment for small-scale village lime-slaking and earth-stabilization works. Tools and equipment for working with and field testing lime-stabilized soil are illustrated in Figures 4.1 to 4.3. The items shown are:

1. Large watertight containers for storing material (e.g. empty oil drums with lids) or storage tanks dug into the ground.
2. Smaller containers – buckets and bowls for immersion testing.
3. Watering can with a rose.
4. A set of sieves or screens with aperture sizes selected as required from those set out in Table 4.1.
5. Hoes, drags, or rakes for lime slaking.
6. Spades and shovels.
7. Wheelbarrows.
8. Boards, sacking, or matting covers or old large sheeting material for shading the lime pits and finished work for curing.

38 BUILDING WITH LIME-STABILIZED SOIL

Figure 4.1 Tools and equipment (1).

9. Timber moulds for brick- and block-making and for sample test cubes and discs and/or mechanical block press.
10. Boards on which to demould and move blocks.
11. Plastic sheets on which to lie blocks and cover them while curing.
12. Lump hammer and/or crushing machine.
13. Plumb bob and level.
14. Personal protective equipment – goggles, gloves, masks, foot protection, etc.
15. Linseed or other oil, turpentine or citrus oil, and beeswax or other wax polish.
16. Clean water for washing off lime.
17. Plastering floats and trowels.
18. Brushes for applying limewash, damping down, and cleaning.
19. Rammers, tamper and pestle for compacting mixes (see Figure 4.3).

FIELD TESTING: STAGE 1 – MATERIALS **39**

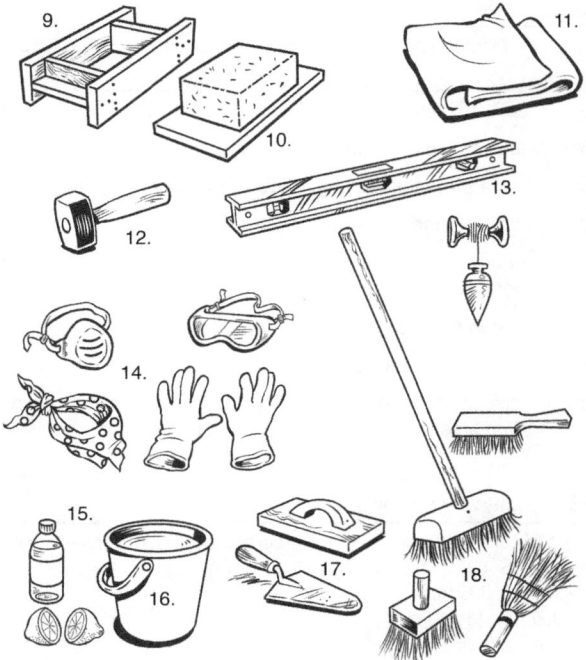

Figure 4.2 Tools and equipment (2).

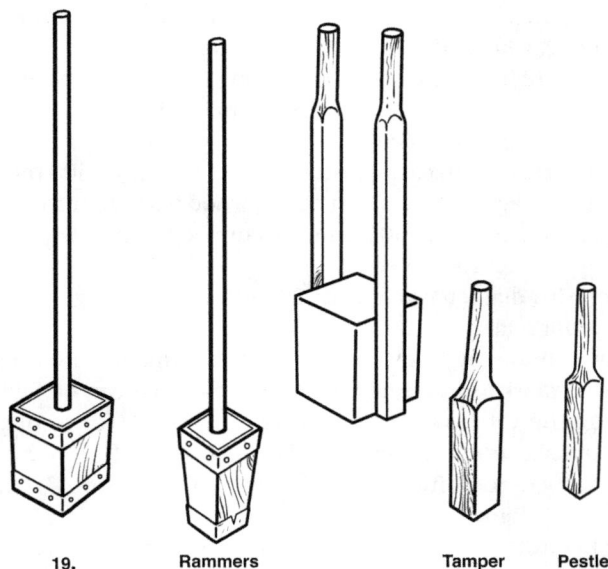

Figure 4.3 Tools and equipment (3).

Table 4.1 Suggested sieve sizes.

ASTM no.	Aperture size/mm	Material particle size established by passing sieve
4	5.00	soil, fine gravel, and coarse sand for the main work; quicklime granules
6	3.35	coarse crushed quicklime for foundations and reactivity tests
8	2.36	lime putty for backing render (levelling) coat, mortar, and foundations
10	2.00	coarse sand, medium soil for field tests
20	0.850	lime putty for finishing coats; quicklime powder for blocks, base render coats, and field tests
30	0.600	medium sand, dry hydrate for coarse work
40	0.450	fine soil for field tests, coarse limewash
80	0.180	coarse pozzolan; finest lime putty (for fine stucco and decorative work); fine limewash; fine sand; dry hydrate for finishing plaster
140	0.106	very fine sand
230	0.063	silt, very fine pozzolan

Note: The above aperture sizes are given for practical field-test purposes but the formal definition of sand to BS 1377-2:1990 is:
- Coarse – 2 mm to 0.6 mm
- Medium – 0.6 mm to 0.2 mm
- Fine – 0.2 mm to 0.06 mm.

Simple, low-cost field-testing equipment.

- Graduated measuring containers: ½-litre/1 pint, 1-litre/2 pints and 3-litre/6 pints. The largest of these must be metal or heat-resistant.
- Reinforcing rod or bar 500 mm (1'8") long and 10 mm (⅜") in diameter.
- 40–50 mm (2") dia. disc test plastic sleeves 10–15 mm deep (⅜" to ⅝").
- Means to construct a drying tunnel or drying oven for soil samples (e.g. clear plastic sheeting).
- Sheet of glass or firm clear plastic approx. 300 mm × 300 mm (12" × 12").
- Linear shrinkage (Alcock's) test box of wood with smooth inner faces and internal dimensions of 600 mm × 40 mm × 40 mm (2' × 1½" × 1½").
- Two contact thermometers.
- Funnel with diameter of at least 200 mm (8") and high sides.
- Waterproof sealant.
- Hollow cylinder or pipe 12.5 mm (½") in diameter with one sealed end, or plastic marker pen sealed flat at one end that can be filled with sand to bring the total weight to exactly 30 g (1 oz) for testing lime-putty consistency using a method based on ASTM C 110.
- Level-topped container approximately 70 mm (or 3") diameter and 40 mm (2") high.
- Lump hammers, roller-pan mixer, jaw crusher and/or other machinery, to crush pozzolan and quicklime to pass the sieve mesh sizes recommended in Table 4.1.

- Permanent markers, masking tape, record sheets, and material for labelling, recording, and monitoring samples.
- Pick axe, for breaking ground.
- 50 mm or 25 mm (2" or 1") high × 75 mm or 100 mm (3" or 4") diameter moulds, for moulding render and screed-mix test samples (could be cut from plastic pipe or bottles).
- 50 mm × 50 mm × 50 mm (2" × 2" × 2") cube moulds (steel, or smooth-sided marine ply or timber).

Mechanized equipment, types and sources. It is often beneficial to conduct research into the availability of equipment, machinery and local manufacturers, not only to assess affordability, but also to establish what materials and skills exist in individual regions. Equipment and tools may be purpose-made or adapted from those produced for other uses, such as agriculture. Typical mechanical equipment and machinery appropriate for small-scale production include block presses like the CINVA ram, mortar mixers, and agricultural backpack sprayers for misting walls (see Appendix 2). Shredding and chopping machines are available for fibre cutting, and a range of crushing and grinding machines and techniques can be adapted for crushing pozzolan or quicklime. Many are in current use for agricultural purposes. (See photo 14.)

4.1.3 Safety precautions

Key points. Lime is an excellent building material. It provides clean and uniform finishes, and can help protect buildings from water – both heavy rains and floods. However, until it is fully cured and set, lime needs to be mixed and handled with care. Lime is a strong alkali and can burn the skin, particularly when in the form of quicklime (burnt limestone), which is activated by moisture. Quicklime in any form, including dust, should therefore not be allowed near or in the eyes, or onto wet or damp skin or clothes, when it would become activated.

- To protect your skin, apply barrier cream, or, if this is not available, any oil (such as linseed or coconut oil) before working with lime, and wear gloves and goggles or glasses.
- Do not work alone when mixing lime. Lumps can explode if only partially immersed. The reaction is 'exothermic', releases great heat, boils and could spit. If it spits into the eye, it can blind.
- Keep a clean source of water close to hand for irrigating the eyes if necessary, for example a hose pipe, or a water bottle that can be used to direct a clean water flow across the eye.
- Crushing quicklime can be dangerous. Wear protective clothing, eye protection, face mask, gloves, long sleeves and shoes or boots (not open sandals) while crushing, and keep others away (see Figure 4.5).
- Do not allow lime, particularly quicklime, to remain on your skin: keep clean water close at hand to wash it off.

42 BUILDING WITH LIME-STABILIZED SOIL

Figure 4.4 Do not splash quicklime onto eyes or skin – it is easy to accidentally splash not only yourself, but also the person next to you.

Slaking. Mixing lumps of quicklime with water to make lime putty is especially hazardous. Large lumps can explode if only partially immersed. The reaction can release great heat, and the mixture can boil and spit. If it spits into the eye, it can blind.

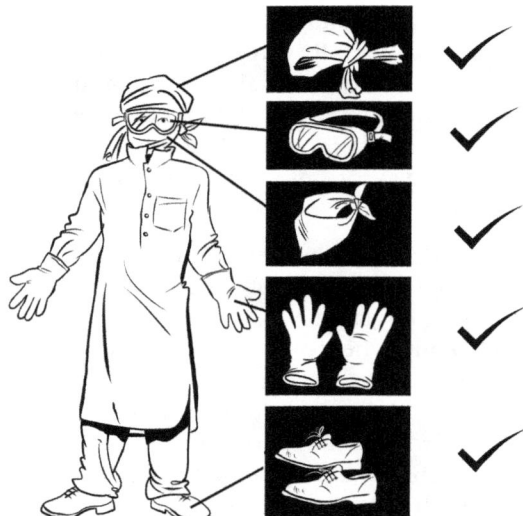

Figure 4.5 Wear protective clothing.

- Break down any lumps that are over 150 mm (6") in diameter before slaking, and make sure there is enough water in the slaking tank to fully submerge them.

- When making putty always add the quicklime to the water (not water to lime) (see Figures 4.6 and 4.7), and ensure that the quicklime lumps are always fully submerged immediately.
- Keep a ready and plentiful water source close to hand throughout the slaking process.

Figures 4.6 and 4.7 For safety reasons, always add quicklime to water, not water to quicklime for making lime putty (unless the water is a well-controlled light and fine spray to produce dry hydrate powder, or the quicklime has first been covered with a soil, sand or other aggregate).

Figure 4.8 Lime putty production – also see Sections 4.1.9 and 4.1.10 for lime-slaking tank and settlement pit construction, and photos 12a and 12b.

44 BUILDING WITH LIME-STABILIZED SOIL

- When slaking, wear eye protection, cover and protect bare skin. Wear long sleeves.
- Wear waterproof gloves and enclosed footwear.
- Keep children and animals away from the lime settlement pit. Surround the pit with some form of fencing or a barrier for safety.

Figure 4.9 Keep children and animals safe by fencing off the lime pit.

Dealing with lime burns. When working with lime, always keep buckets of clean water close by for eye and skin washing.

Figure 4.10 Always have clean water available.

If wet lime splashes into the eye, immediately get help to flush the eye with clean water. Keep flushing the eye for several minutes with running water. If the eye remains bloodshot and sore, seek medical assistance. It is important to wear eye protection to prevent this from happening. Lime, particularly quicklime in the eye, could cause blindness.

Figure 4.11 Remove lime splashes immediately using clean water.

If the skin suffers a lime burn, it will need washing well. The acidity of vinegar or lemon added to the water helps neutralize the lime's alkalinity, as will dabbing the affected area directly with vinegar or lemon juice. If the burn is bad, seek medical assistance (see Figures 4.10 and 4.11).

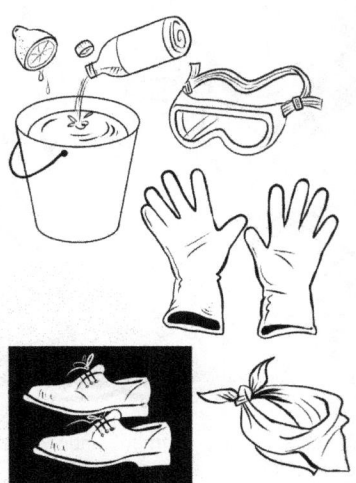

Figure 4.12 Personal protective equipment.

Hot mixes. The term 'hot mix' is used to describe the result of the process of mixing quicklime directly with an aggregate that is mainly soil or sand, which may be wet or damp, or mixed dry with the quicklime, and the mix then dampened all together (see Figure 4.13). To obtain a rapid and even reaction, it is usually helpful to crush the quicklime first, to an appropriate size for the end use of the mix (see also Sections 5.1.2 and 5.6).

In most cases, good-quality, finely powdered quicklime is more reactive than lime putty or dry hydrate powder, and this is particularly beneficial for stabilizing clay-content soils. The powder may have to be produced by crushing or grinding lump quicklime, which can be a hazardous process if not carried out in a controlled way. The safest and most convenient way to use quicklime may therefore be in granular form: that is, particles with a diameter in the region of 5 mm (ASTM sieve No. 4). Granulated quicklime is often available from large manufacturers and producing it on a small scale may not be practical, so crushing the larger lumps to powder may be the best option.

The hot-mix process can be dangerous, as it creates heat and possibly lime dust, which can burn:

- Cover your nose, bare skin and eyes when using quicklime powder and when hot-mixing.
- Check the wind direction and ensure that quicklime dust does not affect anyone or anything downwind of or adjacent to the work.
- The safest way to mix quicklime with an aggregate (sand or soil) is to fully cover the quicklime with damp aggregate first, before adding water.

Figure 4.13 Hot-mixing: add quicklime to a damp mix, add water as required, and allow to slake fully. Mix well, ensure uniformity of colour and keep damp. See Chapters 5 and 6 for variations in application for different building elements.

FIELD TESTING: STAGE 1 – MATERIALS   47

4.1.4 Quality of building limes

Testing building limes. The first stage of testing materials is to examine the quality of potentially suitable building limes.

Simple field tests help determine which form of lime is likely to be satisfactory for stabilizing a particular soil type; it may be in the form of quicklime, dry hydrate or lime putty.

Dry hydrate and lime putty are produced from quicklime, which will need to be of good quality to be effective for stabilizing soil. For best results, the quicklime should be freshly burnt and sufficiently reactive. The reactivity of quicklime diminishes the older it is and the longer it has been exposed to the air. Its quality will also be reduced if it has been under-burnt or over-burnt in the kiln (see Figure 4.19).

In this context, 'lime' means non-hydraulic lime used in the form of non-hydraulic quicklime, non-hydraulic lime putty, or non-hydraulic dry hydrate, unless stated otherwise.

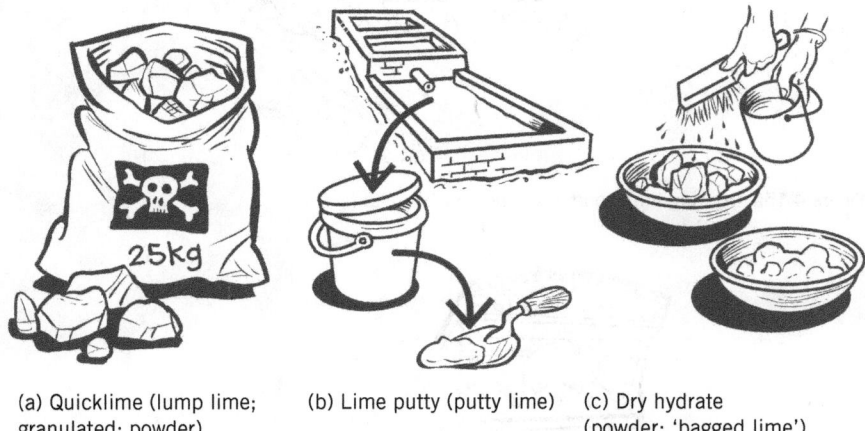

(a) Quicklime (lump lime; granulated; powder)   (b) Lime putty (putty lime)   (c) Dry hydrate (powder; 'bagged lime')

Figure 4.14 Three forms of non-hydraulic lime.

Production of non-hydraulic lime. The lime cycle (see Figure 3.1) starts with the heating (burning) of quarried limestone (calcium carbonate – $CaCO_3$) in a kiln (see photos 8 and 9). The stone must reach a minimum temperature of about 900°C (1,650°F), during which carbon dioxide (CO_2) is driven off to produce quicklime (calcium oxide, CaO). This temperature usually has to be maintained for several days depending on the size of the limestone lumps, efficiency of the kiln and the way it is loaded. Details are given in Michael Wingate's book *Small-scale Lime-burning* (1985).

(a) Non-hydraulic quicklime that is well burnt will often be white or of a light colour and weigh up to 44% less than the original limestone (calcium carbonate, $CaCO_3$) that went into the kiln, due to the loss of carbon dioxide

48 BUILDING WITH LIME-STABILIZED SOIL

Figure 4.15a Burning limestone in a lime kiln, Pakistan.

Figure 4.15b A traditional method of loading quicklime in Southern Pakistan. Bagging before transportation is now best practice.

FIELD TESTING: STAGE 1 – MATERIALS 49

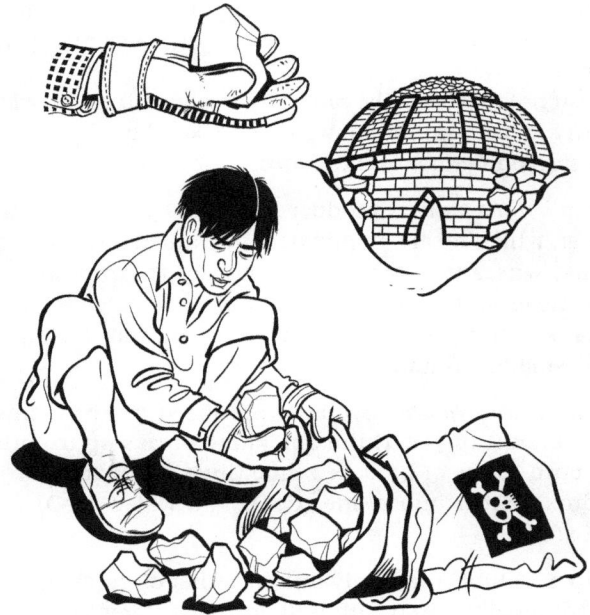

Figure 4.15c Packing quicklime with gloved hands into waterproof and airtight bags.

Figure 4.15d Protect the quicklime from moisture.

(CO_2) during the burn. Fresh quicklime is highly reactive. It must be treated cautiously as it is unstable in the presence of water and will react with moisture very rapidly.

Quicklime, if not used directly, can be hydrated with excess water to make a 'wet' lime putty, or with minimum water to make a 'dry' hydrate powder, and can be safely stored in either of those forms.

(b) Non-hydraulic lime putty is produced by slaking ('hydrating') quicklime in excess water, which means in practice the immersion of the quicklime in a slaking tank, where it is raked continuously to form a slurry, running it through a sieve and letting it settle out as lime putty in a prepared settlement tank. If only small amounts are required, a similar process is carried out in smaller containers.

(c) Non-hydraulic dry hydrate powder is produced by sprinkling quicklime with a small amount of water: the quicklime breaks up to form a powder. Chemically, both forms of hydrated lime (lime putty and dry hydrate) are calcium hydroxide: calcium oxide (CaO) and water (H_2O) form calcium hydroxide ($Ca(OH)_2$).

Carbonation. A lime–sand only mix, whether of quicklime, putty or dry hydrate, is traditionally mixed with well-graded (different-sized particles) and preferably also sharp sand, to form mortars, renders, and plasters. In good curing conditions (shaded and regularly dampened for at least 28 days) these will carbonate on the walls through evaporation of the moisture and re-absorption of the carbon dioxide forming calcium carbonate, which is where the process started:

$$Ca(OH)_2 + CO_2 - H_2O \rightarrow CaCO_3$$

And it doesn't end there ...

Lime takes time. The lime mix increases in strength and protective ability over time as the setting process continues. For this to happen, however, the initial curing process is critical, enabling both carbonation and, for soil stabilization, a chemical (hydraulic) set. The initial curing period is 28 days, during which the work must not be allowed to dry out too quickly so should be kept shaded and regularly dampened (see Section 5.7). Standard compressive strength tests are carried out after the 28 days of curing, but as strength continues to increase slowly, it can be double the 28-day strength after two years.

Carbon sequestration is another advantage. Reabsorption of the carbon dioxide released during the initial heating means the production cycle is almost carbon-neutral (subject to burning method), so lime offers an environmentally sensitive and sustainable construction material. The whole process could be fully carbon-neutral when fossil-free fuels have been sufficiently developed (see Table 1.1 in Section 1.4).

Storage and protection of quicklime. Quicklime should ideally be used fresh from the kiln ('freshly burnt') or preferably within three days of burning.

If it is to be stored or transported, it needs to be sealed in airtight and waterproof bags or containers: quicklime degrades if exposed to the air as it starts to 'air slake' (i.e. carbonate and lose its binding properties), and it is dangerous to allow it to get wet – fire or an explosion could result. Keep airtight, weatherproof bags or containers of quicklime dry and well-sealed until used. Store them on raised ground and protect them from rain. Keep the bags closed tightly. If the age of the quicklime or quality is not known, carry out quicklime reactivity tests, as detailed below, at the time of purchase and then at regular intervals until the quicklime is to be used. Compare the reactivity of this quicklime with fresh quicklime from a local supplier, and recommended field-test results, to ensure consistent quality.

If it is not possible to use freshly burnt (highly reactive) quicklime immediately, or to pack it into completely airtight containers, its quality can be protected and maintained by slaking it as soon as possible and storing it as lime putty under water. The quality of the putty will continue to improve provided that it always has a covering of water. Lime putty may therefore be stored in this way for an indefinite period (see Section 4.1.10).

4.1.5 Preparing quicklime for testing

Safe handling of quicklime. Move quicklime in wheelbarrows, or carry one heavy bag between two people. Do not carry bags of quicklime without adequate skin or head (and ideally eye) protection. Quicklime dust is caustic and can burn in the presence of moisture such as sweat.

Preparation for testing. Testing quicklime reactivity prior to its purchase, delivery and use is essential to ensure that it is of the best quality and is sufficiently reactive. Mixes that incorporate lime of a poor quality are likely to fail.

Methods of field testing quicklime to establish its quality are set out in Section 4.1.6. It is very important that the quicklime is well burnt and contains very little, or preferably no, under-burnt or over-burnt material. Confirm this through testing. If powder is to be used, crush sufficient quicklime for a test sample.

Quicklime can be crushed on a small scale with hand tools and sieved (see Figure 4.16) or on a larger scale with a machine such as a jaw crusher, ball mill, or roller mixer. Small-scale crushing machines are widely manufactured and many hand-, animal- and motor-driven types are available (see Figure 4.17 and appendix 2). The ideal machine to select is one that is able

Figure 4.16 Crushing quicklime to powder by hand and sieving.

to crush both quicklime and pozzolans to the particle sizes recommended in Table 4.1.

- Lime dust is dangerous and should not be breathed in. Wear a dust mask and eye protection while crushing.

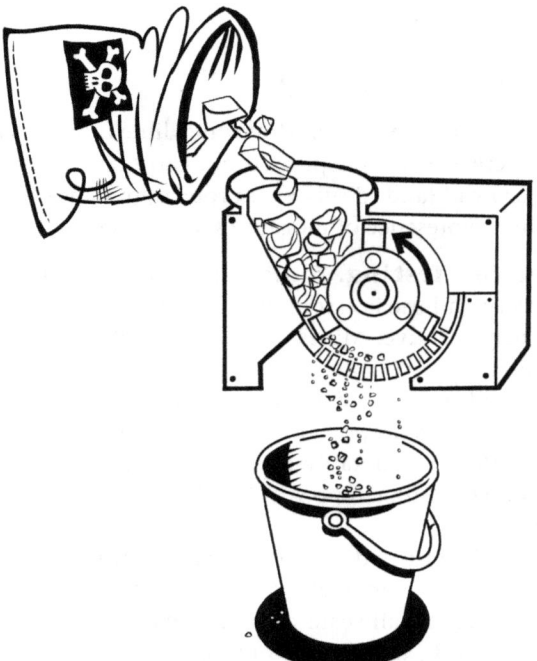

Figure 4.17 Hammer mill – one of various mechanical methods of crushing quicklime.

- Wear gloves and protective clothing (see Figure 4.5).
- Keep children and animals away from quicklime crushing, hot-mixing and lime slaking (see Figure 4.18).

The powdered quicklime used should be fully reactive and pass through a 0.850 mm (ASTM No. 20) sieve for blocks, or a 0.180 mm (ASTM No. 80) sieve for first render and plaster coats (see Table 4.1). Well slaked lime putty, however, is often best for renders and plasters, particularly finishing coats, unless best quality, finely powdered quicklime is available.

Figure 4.18 Keep children and animals safely away from all lime work.

4.1.6 Field testing quicklime

It is important to use lime that is fully reactive. This section describes field tests to determine whether quicklime has been correctly burnt and is sufficiently reactive.

Observation tests. An initial assessment of the quality of quicklime can be made on the basis of observation before proceeding to more complex field tests.

Under-burnt limestone will be heavier than a fully burnt stone of the same size. It may contain a core of stone which has not fully calcined because it has not been exposed to sufficient heat. The whole stone may have a recognizably different colour, texture, and density, or it may be just the core that differs from the surrounding quicklime. An under-burnt core would remain in the water as residue following slaking (see Figure 4.19). To check whether two stones have been under-burnt, knock the cores together after removing the surrounding quicklime. Wear gloves. If they have remained as stone, they will sound and feel like two stones clacking together, which is what they are.

It is important to avoid crushing under-burnt material and including it with good, burnt lime, as it is not sufficiently reactive, does not have the ability to stabilize the soil, and will have very few binding properties, if any.

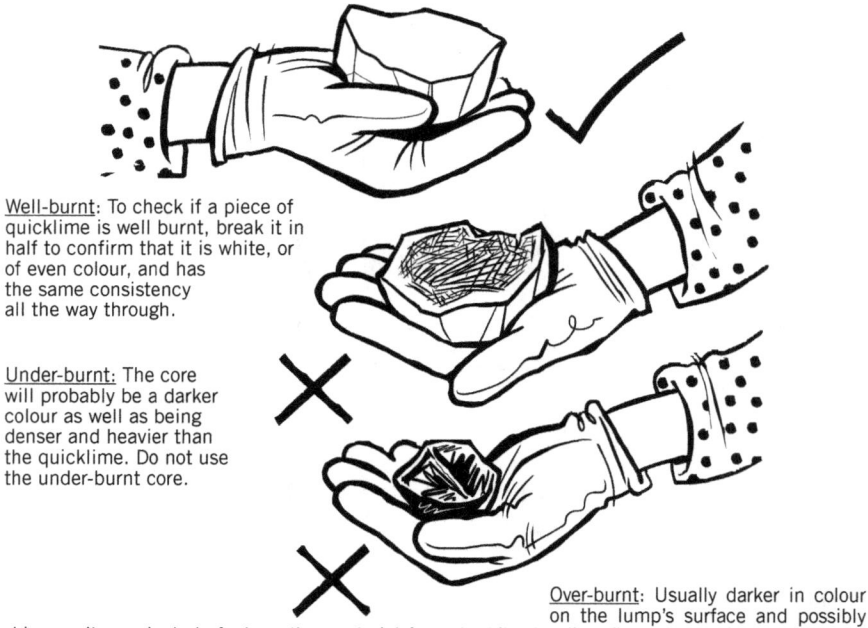

Well-burnt: To check if a piece of quicklime is well burnt, break it in half to confirm that it is white, or of even colour, and has the same consistency all the way through.

Under-burnt: The core will probably be a darker colour as well as being denser and heavier than the quicklime. Do not use the under-burnt core.

Over-burnt: Usually darker in colour on the lump's surface and possibly shiny, as it may include fuel or other material from the kiln that liquefied in excessive heat. This is likely to have left clinker deposits which set hard on cooling. Do not use clinker or over-burnt stone.

Figure 4.19 Check the quicklime is not under-burnt or over-burnt. If there is any poorly burnt limestone, separate it out. Only use well-burnt lime.

The method of burning, kiln design, fuel used, and the type and size of limestone lumps will all affect the quality of quicklime produced. The burning temperature of external uncovered small-scale kilns will be affected by weather conditions, particularly wind strength and direction. Lime should be burnt at about or a little above 900°C (1,650°F).

Over-burning is likely to be minimal if wood is used as fuel, but extremely high temperatures (1,400°C/2,550°F or more) can be reached with coal and coke. Generally, the higher the temperature and the longer this has been maintained, the greater the amount of over-burnt material and clinker that is likely to be produced. Pure lime can become dead-burnt and lose its reactivity, and hard-burnt particles may take weeks or months to slake. Over-burnt material can often be recognized by the presence of a darker hard crust or clinker on the surface, or surrounding the burnt stone.

In addition to examining the kiln product, a further quality check is to compare its reactivity with well-burnt limestone. The field tests described below are ways to ensure quicklime has been fully calcined and is fresh, without having absorbed detrimental amounts of carbon dioxide or moisture.

Six-second test. Place small lumps of quicklime (about 25 mm or 1" diameter) into the bottom of an open mesh container such as a small metal sieve or kitchen colander. Dip the container and contents into a bucket filled with fresh clean water, holding the container so that the quicklime is below the water for six seconds exactly. Then lift out the container, allow it to drain, and empty the contents onto a dry inert surface such as a metal plate, bowl or stone for observation (see Figure 4.20). Good-quality lime will behave in one of the following ways:

Pure (non-hydraulic) lime: Good-quality pure (non-hydraulic) lime hisses, swells rapidly, breaks up, increases in temperature sufficiently to produce water vapour, and turns to powder almost immediately, or within a few minutes. This process of turning quicklime to powder is called slaking to dry hydrate of lime.

Figure 4.20 Six-second test.

56 BUILDING WITH LIME-STABILIZED SOIL

If put into excess water, the dry hydrate turns to putty, the consistency of which remains unchanged over time while under water (it does not harden). Unlike a hydraulic lime, a non-hydraulic lime will not set under water on its own. The volume of a non-hydraulic slaked putty or dry hydrate is usually at least double that of the quicklime.

Natural hydraulic lime: Limestones that contain a sufficient amount of certain minerals, mainly silicas and aluminas, will produce hydraulic lime when burnt. In the six-second test, hydraulic lime expands and breaks down to powder (slakes) more slowly than a pure lime. The most eminently hydraulic limes take the longest to slake. If the properties of the original stone are not known (i.e. whether the stone is non-hydraulic or hydraulic), a six-second test where the lime takes a long time to break down may evidence either a poor-quality (lean) non-hydraulic quicklime, or a quicklime with some hydraulicity. To check if the quicklime could be hydraulic, further test it by placing it under water to see whether it will set on its own, or when mixed with sand. Setting would confirm it to be a hydraulic lime. The degree of hydraulicity is related to the time it takes to set solid, generally between 3 and 20 days (see Holmes and Wingate, 2002: Appendix 1). The most hydraulic limes (eminently hydraulic) take the longest to slake.

Gain-in-weight measurement. The process of burning can reduce the weight of pure limestone by up to 44%. In reabsorbing the carbon dioxide as well as moisture from the air, quicklime gains weight and loses reactivity (see Figure 4.21). This can also be an indication of the extent of undesirable air slaking as well as water absorbed.

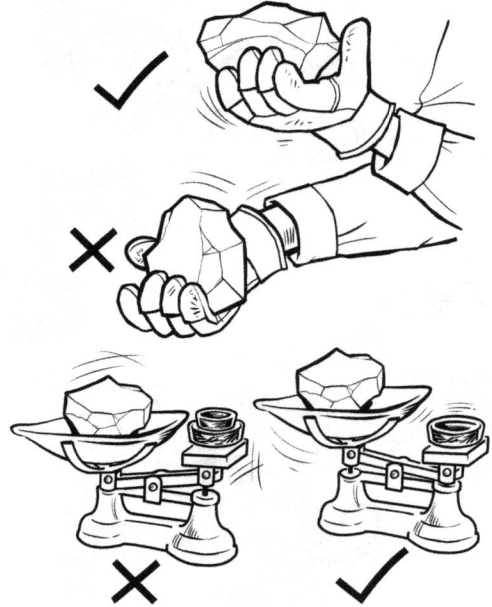

Figure 4.21 Weigh the quicklime to check that it has lost enough weight to be fully burnt.

A sample of the quicklime under test may therefore be carefully dried and weighed for comparison with a sample of fresh, well-burnt quicklime of identical source and volume. The initial control sample must be taken fresh from the kiln immediately following a good burn. A good burn can be judged by the successful conversion of all, or at least 95%, of the stone in the batch (i.e. no over-burnt or under-burnt material), and by it all slaking thoroughly, leaving no residue. The difference shown by the increased weight of the test sample over the control sample, will be that of the moisture and carbon dioxide absorbed.

The amount of carbon dioxide alone may be checked in the field. Heat the sample (over a fire or in the sun) to, say, 80°C (176°F) for half an hour to drive off the moisture, and re-weigh it. Crush and sieve the quicklime samples to between 3 mm and 5 mm particle size (⅛" to ³⁄₃₆" – ASTM sieve sizes No. 4 and No. 6), then dry and weigh 1 litre (2 pints) of each of the control sample and the test sample for comparison. The reduction in weight of limestone when all carbon dioxide has been driven off should be in the order of 44% (Wingate, 1985). As a guide, specifications for commercial lime in the United Kingdom state that the carbon dioxide content should not exceed 5%.

Quicklime reactivity. There are two versions of this comparative test based on measuring the amount and rate at which heat is produced by the chemical reaction of calcium oxide with water. One method, should the water boil, is timing how long it takes to reach boiling point, and the other is to measure the rate of the water's temperature rise with a thermometer.

Fully calcined, freshly burnt, pure (non-hydraulic) quicklime will rapidly increase the temperature of water in which it is immersed. The test is therefore more appropriate for the purer limes that are mostly used for soil stabilization.

(a) The first version, a *boiling time measurement*, is carried out as described below. When carrying out the test, wear safety goggles and stand well back as the mixture may bubble and spit when boiling.

1. Pour 1 litre (2 pints) of water at a set temperature, which could be room temperature, or about 25°C (77°F), into a 3-litre (6-pint) metal container. This is a comparative test so it is important that the temperature of the water at the point the quicklime is immersed is the same for every test.
2. Fully immerse half a litre (1 pint) of a representative sample of quicklime. All quicklime lumps in the sample need to be the same size. Although they could be about 10 mm (about ½") in diameter, it is better to use quicklime crushed to powder (and, for accuracy, sieved through a 3.35 mm (⅛"– ASTM No. 6) sieve). (If not using metric measurements, it is possible to simply transpose pints for litres – e.g. pour 1 pint of room-temperature water into a three-pint metal container and immerse half a pint of crushed or powdered quicklime.)
3. Record the exact time taken for the water to be brought to the boil from the moment of immersion. This can occur in less than a minute with some limes, and instantaneously with very fresh lime. Experience has shown that in ambient temperatures of 25°C to 30°C reasonably reactive crushed quicklime should boil water in between one and five minutes.

Test quicklime produced at each firing, or from different kilns (different suppliers and distributors).

- If the quality of quicklime in each test is consistent, the time taken to raise the same volume of water from the same starting temperature to boiling point should also be consistent.
- If it takes longer, or does not boil at all, this indicates that the quicklime is less reactive, possibly due to poor burning or having been left exposed to the air for too long. Alternatively, the quicklime may have been prepared from stone taken from a different bed in the quarry that contains a lower proportion of calcium carbonate, and it may be too lean, or even hydraulic.

Some pure (non-hydraulic) lime samples may cause the water temperature to come close to but not reach 100°C. It would be advisable to avoid using such lime for building purposes as it may not be sufficiently reactive. However, it does not have to be wasted: it could (after slaking) be used as agricultural lime to raise a soil's pH.

Figure 4.22 Lime reactivity test – boiling time.
Stand well back as the mixture may bubble and spit when boiling. Wear safety glasses.
1. Pour 1 litre of room-temperature water into a 3-litre metal container.
2. Add 500 ml of crushed and sieved (3.35 mm) quicklime to the water in the metal container.
3. If the lime is of reasonable quality it will boil the water within five minutes.

It may be worth checking if the lime sample is hydraulic by carrying out a field test described above on an identical sample from the same batch. The sample could be left under water either on its own and/or with sand, to check if it will set, and the second reactivity test sequence described below could be conducted.

(b) The second, a *temperature measurement method*, is a comparative test which is more precise and requires a thermometer and exact timing and recording. It can be useful not just for comparing batches of non-hydraulic limes, but also for fresh, feebly hydraulic limes which do not bring water to boiling point, and for eminently hydraulic limes where the temperature rise may be very small and much slower.

Use the same sample volumes and equipment as above and follow the same initial procedure, but after immersing the quicklime sample use a thermometer to record the temperature every 30 seconds (see Figure 4.22).

The rate of temperature rise (the difference between consecutive readings) can be compared to that over previous intervals, or related to experience with other limes. Temperature changes for the slower slaking hydraulic limes can be slight and are more difficult to record. To make measurement easier, the container should be well insulated, or a vacuum flask could be used.

4.1.7 Dry hydrate preparation

Dry hydrate is prepared by sprinkling small amounts of water on reactive (tested) lumps of well-burnt fresh quicklime. For testing and small-scale use, this could be done using a brush or watering can. The lumps of quicklime should soon start to 'bloom' as they swell and crumble into powder (see Figure 4.23). It may take a few minutes or some hours for all the lumps to become powder, depending on both the amount of quicklime to be slaked and its reactivity. For a pure

Figure 4.23 Preparing dry hydrate by hand.
In a metal container, sprinkle water over a few quicklime lumps. If the quicklime is good quality, it will soon break down into powder. Sieve the powder before use.

60   BUILDING WITH LIME-STABILIZED SOIL

(non-hydraulic) lime, the quicker it breaks down, the better the quality. Heat will be given off. Sieve the resultant powder through a 0.6 mm (ASTM No. 30) sieve, and the freshly prepared dry hydrate is ready to use. Alternatively, pack it in well-sealed and fully airtight bags or containers on the same day it is prepared.

There are simple ways of mechanizing this process if demand is sufficient, and fully automated equipment has been developed to hydrate, crush, sieve, and bag lime on a large scale.

Dry hydrate powder and quicklime powder: It is difficult to tell the difference between dry hydrate powder and air-slaked lime, both of which can be bagged and sold in powder form. Air-slaked lime should not be used. It is a better policy therefore to purchase (tested) lime as quicklime lumps in airtight sealed bags fresh from the kiln, rather than in powder form, unless you know for certain that it will be best quality, fresh, dry hydrate, and that it too is packed in well-sealed airtight bags.

The quality of quicklime powder can be checked by using a gain-in-weight measurement field test (see above) to determine the amount of carbon dioxide that has been re-absorbed. Air-slaked quicklime powder will have lost some or all of its reactivity, depending on the length of time it has been exposed to the air. If the extent to which reactivity has been lost means it is no longer suitable for its intended use, it should be discarded.

4.1.8 Field testing dry hydrate

Fineness. Sieve testing will give an initial indication of the quality of a dry hydrate. If production, packaging, transport, and storage have been in accordance with the recommended national standards, the lime should pass simple particle-size tests.

Standards usually require the majority (99%) of all hydrate to pass through a 180 micron mesh (ASTM No. 80) sieve. The degree of fineness of about 0.2 mm is advisable for fine plastering but is not necessary for the majority of other applications for which building limes and stabilized soils are required. The hydrate will normally be acceptable in terms of fineness if 100 g (3.5 ounces) passes a 0.85 mm (ASTM No. 20) sieve after five minutes of continuous sifting, leaving no residue. The sieving should be carried out by shaking without brushing, rubbing, or punching the lime through.

Density. The density of dry hydrate may be determined in the same way as described for determining that of lime putty in Section 4.1.11: that is, by using a jug which holds exactly 1 litre as a density vessel. Some national standards give a range of maximum density figures for each class of building lime. The ASTM guideline value for maximum bulk density of pure dry (non-hydraulic) hydrated lime is 0.5 g/ml (or 1 litre to weigh 0.5 kg).

Dry hydrate from hydraulic lime. As detailed earlier, natural hydraulic limes are made by burning limestones which contain active clay. The active clays in the limestone combine with lime when the stone is burnt, to produce a hydraulic lime which will set under water. Commercial natural hydraulic lime is usually sold as dry hydrate in airtight bags and was traditionally classified as feebly, moderately, or eminently hydraulic. These definitions were widely used in the past and continue to be used in many countries. They do not, however, relate to the current BS EN (European) Standard for Building Limes, which have a different classification.

Provided the source limestone containing active clay is of a consistent mineralogy and has been well burnt, the bulk density of its dry hydrate can give an indication of its level of hydraulic content. The ASTM bulk density levels for each lime classification are given in Table 4.2.

Table 4.2 ASTM bulk density levels for hydrated lime classification (both non-hydraulic and hydraulic dry hydrates).

Dry hydrate of lime	Bulk density (g/ml)
White (pure) lime – non-hydraulic	0.5
Feebly (slightly) hydraulic lime	0.65
Moderately hydraulic lime	0.65–0.8
Eminently hydraulic lime	0.9–1.0

4.1.9 Lime putty slaking tank and settlement pit construction

Lime putty is generally the preferred form of lime for lime and sand only mixes, and for lime-stabilized soil plasters and renders when best quality, fine quicklime powder is difficult to produce. This section describes how to produce and store lime putty on a small scale, particularly in remote rural areas and in the village environment.

Figure 4.24 Preparing the lime pit. For safety, always keep children and animals away from the lime-slaking and settlement pits.

Choose and prepare the site. The site for lime slaking and size of the settlement pit will vary depending on the amount of putty required at any one time in the building programme. Choose a site with good transport access, where water is readily available and from which children and animals can be kept away. Select a location with elevated, shaded ground for storing bags of quicklime. It is helpful to construct the slaking and settlement tanks (or pits) on a slope in the ground, with the slaking tank immediately above the settlement pit. This allows gravity to assist the putty during slaking (often referred to at this stage as 'lime slurry' or 'milk of lime') in flowing easily from the slaking tank to the putty storage pit below. Allow enough space for both a shallow (0.5 m (1' 9") deep) lime-slaking tank of at least 1.5 m × 0.9 m (5' × 3') and one putty settlement pit, approximately 0.9 m (3') deep and measuring at least 1.5 m × 2.5 m (5' × 8'), directly adjacent to and below it. These are minimum dimensions for permanent use, but two half oil drums, or similarly sized containers, may be adequate for small or temporary putty production (See photos 12a and 12b).

Build an adjoining slaking tank and settlement pit. The tank for slaking the quicklime needs to be behind and higher than the second tank (or pit), in which the lime putty is allowed to settle out. Both should be dug out, or built up from blocks and mortar of the same strong hydraulic mix, and the sides and bases should be plastered with similar strong mixes.

In the higher *slaking tank*, leave an opening about 300 mm (12") above the base at the end for a weir and chute – or drainpipe – over the settlement pit, so that lime slurry can run through a sieve placed underneath the chute, directly into the pit below.

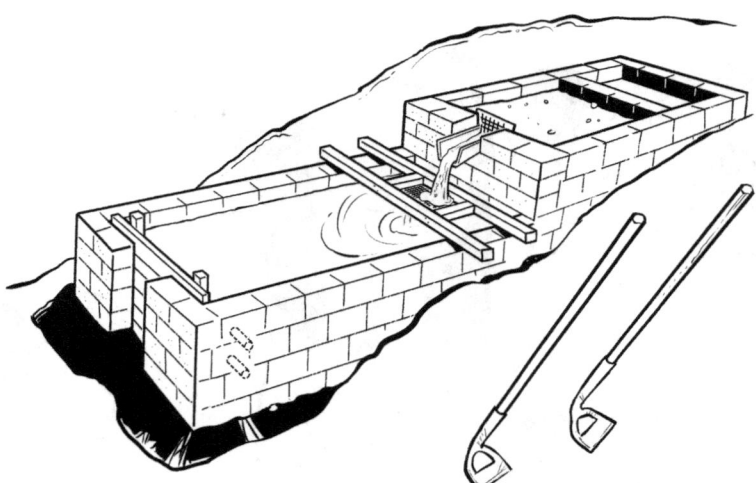

Figure 4.25 Slaking tank and settlement pit for producing lime putty, and long-handled hoes for safe slaking.

FIELD TESTING: STAGE 1 – MATERIALS   63

A small drain-off pipe can also be located below the chute at the floor level of the slaking tank. It must be well sealed during the slaking process but is useful for draining the tank before cleaning. This enables defective material (under-burnt or over-burnt stone) to be gathered and weighed when all slaking has been completed, in order to establish the precise quality of the purchased quicklime. (See 'settlement tank drain off pipes' in Figure 4.27.)

Figure 4.26 Lime slaking to putty in tanks. Keep agitating the quicklime lumps at the bottom of the slaking tank until they have dispersed fully. Overflow pipes should be built into both the tank and the pit to allow excess water to be drained off easily.

Sieves for above the settlement tank, through which the milk of lime will flow, may be 5 mm or 3 mm (ASTM No. 4 and No. 6) or even finer ($3/16''$ down), depending on the intended use of the putty (Table 4.1 gives further detail). Sieves will keep out smaller lumps and unslaked material from the settling putty in the settlement pit, which could damage the finished work.

Settlement pits for settling out the lime putty will vary in size subject to the requirements of each project, but initially allow for a pit of the dimensions described above and allow space for future expansion. A sloping ground surface, and sealed shuttering or hatch at one end, would assist access to the settlement pit for ease of putty removal (see Figures 4.25 and 4.8). The settlement pit is where the lime

64 BUILDING WITH LIME-STABILIZED SOIL

putty can be kept stored under water, maturing (and improving in quality) for weeks or months. It is an effective method of storing fresh lime if it cannot be used soon after being burnt. Cover the pit to protect it from direct sun. This helps to keep the putty clean and to prevent water from evaporating too quickly (see Figure 4.30).

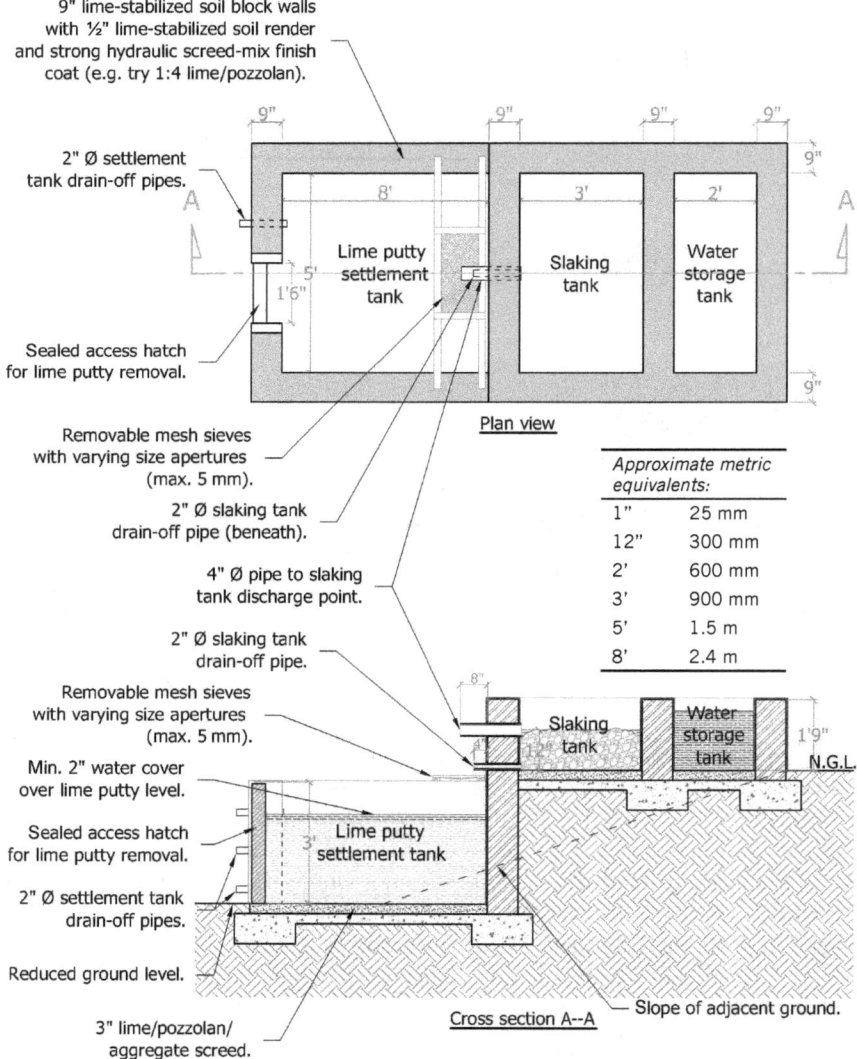

Figure 4.27 Slaking tank and settlement pit dimensions (not to scale). Minimum sizes are given for manual slaking by two to four people. The putty settlement pit may be increased in size if needed, subject to the extent of construction for which the lime is required at any one time.

Ensure that the site is safe and secure at all times. Where applicable, fence the site off to eliminate the risk of falling into the pit, especially for children and animals.

Alternative methods for lining pits. The slaking tank and settlement pit sides should be sealed or lined to help hold water for longer, enable regular use, and keep the putty clean. This can be done in various ways, subject to local conditions. Methods to consider include the following:

- Line the sides and base with well-burnt clay bricks bedded in hydraulic lime mortar or lime and pozzolan. If all the bricks are well laid with a smooth surface finish (fair face), there may be no need to render. Soft burnt bricks should not be used.
- Line the sides and base with lime-stabilized (unburnt) blocks or bricks, and hydraulic lime mortar or lime-stabilized soil mortar. There may be no need to render the tank and pit walls provided they are well constructed, with well-compacted blocks and a level finish. A hydraulic lime (or lime-stabilized soil) render or floor-screed mix could give additional durability. (See Chapter 6 for suggested block, mortar, render, and floor-screed trial mixes.)
- Finish the floor with a screed mix of powdered quicklime and finely sieved pozzolan, or a mix of lime, pozzolan, and well-graded coarse sand. (See also Section 6.11 on strong hydraulic mixes for floor screeds.)
- Add waste marble dust or other available hard granular material, such as crushed limestone grit, to the render mix for the floor, and possibly the sides also, to improve their wearing qualities.
- Use hydraulic bedding mortar and grout with smooth-surfaced and finely jointed stone or other hard material to make a level and durable finish.

Rendering the sides and base. Trowel a finish render onto well-keyed and wetted pit walls and possibly also the base, in coats 10–20 mm (about ½–¾″) thick. A second render coat with fine aggregate and pozzolan may increase the impact resistance and durability of the lining and will improve the water-holding ability of the tank or pit and help keep the putty clean. If the slaking tank is intended for frequent and constant use, it would be advisable to have a hardwearing and smooth floor finish rather than a render or screed (see above). Cover the pit after rendering, damping down, and during and after curing as detailed below.

Protection for curing. It is important that all lime-stabilized mortar, render and floor screeds are given time to cure and harden before the tank is used for slaking. Mortar and render will need to be kept damp and protected from direct sun or heavy rain. The hydraulic set will be at its most effective and will continue to increase the longer the work is kept damp. Keep the pit covered with wetted sacks, cloths, plastic, or a weighted lid, and keep damping down for about four weeks before using it for slaking and putty production.

Figure 4.28 Plastering the sides of the lime pit.

4.1.10 Lime putty production

The following procedure describes the general principles of good-quality putty production, and is of particular relevance for the method of slaking and settlement shown in Figures 4.8 and 4.26.

Slaking. Hydrating and mixing the quicklime with excess water without machinery is active work. An exothermic reaction is created, causing a lot of heat, so it may be better to slake at a cool time of day. Half fill the shallow slaking tank with water. Following this, one person carefully empties half a 25 kg bag (half a 50 lb bag) of fresh quicklime into the water without splashing, while another person keeps raking the mix with a long-handled hoe. Make sure the quicklime lumps are always completely covered with water, as they are more likely to explode if they are not fully immersed. The mix will get very hot and can boil and spit, so it is important to wear protective clothing. Continue to add more water, and no more than half a bag of quicklime at a time. It may take several hours for a team of three or four people to fill the putty storage tank, so shade the area well and avoid working in the hottest part of the day.

Moving the slaking quicklime and putty around with the hoes also stops it forming sticky lumps at the bottom of the tank. A simple method is for two people with hoes to gently rake the mix continuously whilst another ensures the quicklime is covered with water at all times. As the slaking tank is filled with more quicklime and water, the resulting lime slurry will flow down the chute and through the removable sieve into the lime settlement pit below. Any large under-burnt or over-burnt lumps will stay in the slaking tank. Smaller debris, slow slaking and under-burnt particles will be retained on the sieve.

Excess water on top of the putty in the lime settlement pit can be recycled during the slaking process by using buckets to carry it back up to the slaking tank. (See photos 12a and 12b.)

FIELD TESTING: STAGE 1 – MATERIALS 67

Figure 4.29 Use long-handled hoes to keep the quicklime lumps moving and to ensure all quicklime is fully submerged under water at all times.

Safety. Always have clean water ready for washing off lime when slaking. Keep a container of clean water next to the work area for washing eyes or skin. Keep open bags of quicklime covered, and keep them a safe distance from the water supply and slaking pit, to avoid the risk of water splashing on them.

Settling. Over a period of days, for up to two weeks or more, the lime will continue to absorb water, expand, and settle to the bottom of the settlement pit as a thick lime putty. Only lime putty of the appropriate density as described above should be used for most purposes, particularly for lime-stabilized mixes. This settling period will vary depending on the limestone type, and generally the purer the quicklime, the quicker the required density can be achieved. It is important that sufficient slaking time is given to produce good-quality putty of sufficient density (see Section 4.1.11 for details on field testing lime putty density).

Storage. Keep the lime putty in the settlement pit covered with at least 25 mm (1″) of water. If the lime starts to dry out, it can harden, carbonate and become unusable. Check regularly for evaporation, and top up the water when needed.

After slaking, as well as keeping water over the putty, keep the top of the pit well covered, preferably with boards and a tarpaulin, for both safety and to keep the putty clean. Keep the putty settlement pit shaded from direct sunlight, which helps to minimize evaporation and the risk of the putty

Figure 4.30 Keep the lime putty under no less than an inch of water at all times and cover the settlement tank to keep the putty clean and the area safe.

drying out. In due course, if appropriate and if large quantities of putty are required, the putty can be removed and stored in sealed or covered containers in a safe and shaded store provided that it is not allowed to dry out.

Maturing. Leave the lime putty in the settlement pit (under no less than 25 mm (1″) of water) for as long as possible – at least two weeks, and preferably four weeks or more. The longer it matures under water, the better the quality. Not all the lime fully slakes and absorbs the water immediately. Lime takes time to do this. This has been called fattening-up, which is the process of CaO absorbing water (H_2O) and becoming $Ca(OH)_2$ i.e. calcium oxide becoming calcium hydroxide. After this, and over a long time, the putty continues to improve due to changes in the minute calcite crystals.

4.1.11 Field testing lime putty

To achieve a hydraulic set for stabilizing soil, particularly for flood-resilient mixes, it is essential that all lime putty is tested for quality before use. If the lime putty is too thin, it may not stabilize a clay soil, as there may be insufficient lime to create a hydraulic reaction with active clay or pozzolan.

Before taking lime putty from the settlement tank, for testing or use, siphon off or drain all the water from the top. Only use the thicker and denser putty which settles out below. Avoid mixing water back into the putty: the putty used for the main work must be the same density as used in the tests.

Density. Several national standards set an upper limit of 1.45 g/ml (1.8 lbs per pint) for the density of lime putty of standard consistency. The putty's density (specific gravity) is calculated by dividing the maximum mass of the putty in grams by its volume in millilitres. A field-test method is to use any container that holds exactly 1 litre (2 pints). This can be simplified for a field test by using a jug that will hold exactly 1 litre and no more when filled to the brim, with the top of the jug kept level throughout the test.

1. Weigh the container.
2. Fill the container with the putty and ensure all air is expelled by tapping it down. Carefully strike off surplus from the top then weigh the container and putty.
3. Continue to add putty and tap down until there is no increase in mass, strike off, and weigh.
4. Deduct the mass (weight) of the container to find the maximum mass of putty. This is the same figure as the density, calculated in this test as weight. It should be close to 1.45 kg per litre or 1.8 lbs per pint (see Figure 4.31).

Figure 4.31 Lime putty density testing.

70 BUILDING WITH LIME-STABILIZED SOIL

All lime putty needs to be of the appropriate density. If it is not, the density should be adjusted. If 1 litre of putty weighs only 1.35 kg (or if 1 pint of putty weighs less than 1.8 lbs), it is starting to get too thin and its binding properties will be reduced. If the putty has been taken from the settlement tank too soon or water on the top has not been drained off, it is likely to be too wet. Allow the putty to settle out in the tank for a few more days and make sure all covering water is drained off before removing it and testing for density again. If it is too dense, add a small amount of water and mix it in well and re-test until the correct density is achieved.

Consistency. If there are no scales available, a basic field test to check whether putty consistency is adequate for good workability and binding purposes is to fill a container 75 mm (3″) in diameter and 50 mm (2″) high with putty, in a

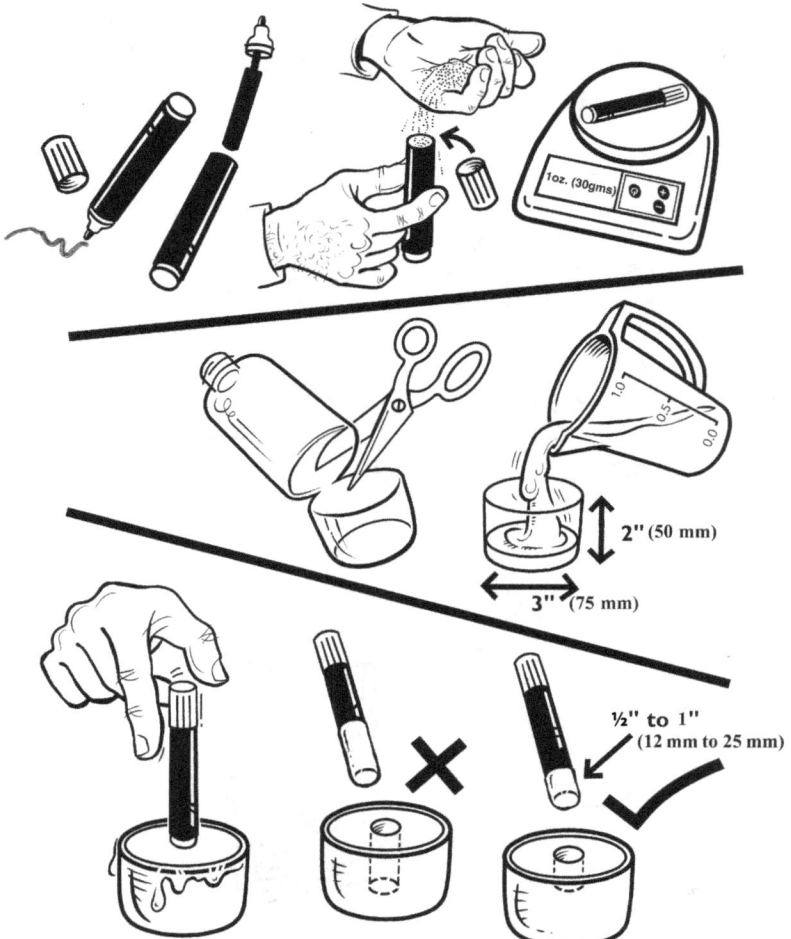

Figure 4.32 Lime putty consistency test.

similar way to filling the container described above. Place a 30 g (or 1 ounce) 12.5 mm (½") diameter plunger on the surface of the putty at the centre. The putty consistency is considered satisfactory if the plunger sinks under its own weight to a depth of between 12.5 mm (½") and 25 mm (1") within 30 seconds. A suitable plunger might be made from a whiteboard marker pen of the same diameter, emptied and filled with sand until it weighs 30 g (see Figure 4.32). Alternatively, a hollow pipe, also of the same diameter, could be filled with the correct amount of sand to bring it to 30 g and sealed at both ends.

Soundness. A test to ensure that lime putty is sound, and to establish its quality, is particularly important if it is to be used for fine finishes such as internal plasterwork, limewash, and decorative modelling.

Spread a thin layer of putty, about 2 mm (¹⁄₁₆") thick, on a sheet of glass or clear plastic and hold it in front of a strong light. If dark spots can be seen, there may be over-burnt or unreactive material that could cause defects at a later date, particularly to finishes, due to delayed slaking. Select only best quality putty, particularly for finishes; if best quality is not available, unslaked particles must be sieved out.

Fineness. Several national standards suggest that all putty for mortar and render base coats should pass through a 2.36 mm (ASTM No. 8) sieve, and, for finishing coats, a 0.85 mm (ASTM No. 20) sieve, leaving no residue (ASTM C5-79). If the putty is to be used for work of the highest standard, such as internal plaster or decorative stucco, it should pass through a 0.18 mm (ASTM No. 80) sieve. In order to achieve this finer material, the putty, or milk of lime, should be washed through the sieve in a diluted form and then allowed to settle out as putty again. This can be done by selecting the appropriate sieve size when slaking (see Table 4.1).

4.2 Soils

Reference here to soils for building purposes will always mean subsoil – after the removal of all topsoil and organic matter.

There are many types of soil, so it is essential to investigate and test them in Stage 1 of the field-test programme. Assessing a soil's composition and mineralogy is a prerequisite to formulating the optimum mix, which may require further modification and stabilization of some soils for specific building elements.

4.2.1 Introduction to soils

Soils are formed mainly through the weathering of rocks over millions of years. They will vary according to the types of rock from which they have been eroded and the minerals they contain. The main ingredients of soils are stone, gravel, sand, silt, and clay. It is rare to find a soil with ideal building properties, although some soil types are naturally suited to particular construction methods or building elements. It is therefore important to check that there are appropriate proportions of clay, silt, sand, and possibly gravel in the soil, to enable stabilization with lime, and to check if the soil needs modification for stabilizing a wide range of mixes for different building elements that are to remain durable in wet conditions.

4.2.2 Ingredients of subsoil

Gravel. (Particle size: 20 mm to 5 mm, ¾" to ¼".) Definitions vary, but for these field-testing purposes, small gravel with particles of 5 mm to 2 mm (ASTM sieves No. 4 to No. 10) is treated as sand. Gravel can be described as consisting of small grains of rock. When mixed with lime and sand, it can form lime concrete for bulkier building elements. Gravel helps to give compressive strength and reduces shrinkage of the building material. Appropriately sized gravels can be selected for mixes for larger elements, such as well-compacted trench footings and floor slabs. Gravel can occur naturally, but it may also be produced artificially by crushing rock. Crushed rock is widely available, particularly where it is also being used on a large scale for road construction and concrete aggregate.

Sand. (Particle size: 5 mm to 0.106 mm, ASTM sieve sizes No. 4 to No. 140.) Grains of sharp, coarse, and well-graded sand provide the skeleton of a building material, give it strength, and reduce shrinkage. Many hill sands are sharp and angular, as are many high river sands. Lower river sands, seashore sands, and desert sands often have rounded particles and feel 'soft' because the sharp edges have been eroded by water or wind. A combination of large-grained (coarse) and small-grained (fine) sharp, angular sand helps to produce a strong bond in mixes. A sandy soil feels grainy and will not stick together when wetted.

Silt. (Particle size: 0.063 mm to 0.002 mm, ASTM sieve size No. 230 and below.) Silt consists of tiny particles, most of which are too small to see with the naked eye. Silty soils feel silky, and the particles are much smaller than sand. A very silty soil will need the addition of clay or sand or both for use as a building material. It is recommended that a soil's silt content does not exceed 20% for modified and stabilized soil mixes, and 6% for lime–sand (and lime–sand–pozzolan) mixes.

Clay. (Particle size: 0.002 mm and below.) Clays consist of miniscule particles, smaller than 0.002 mm, which are too small to see with the naked eye. These particles are chemically different to other grains in the subsoil and can often swell when wet and shrink when dry. Clay is sticky, creates a binding force, and can bind other particles together into useable building material. There are numerous types of clay, many of which are abundant and suitable for stabilization with lime. The best soils to work with lime for stabilization will contain significant proportions of both clay and sand.

Silty, sandy and gravelly soils do not bind. They need clay or lime or both as the binder to make a cohesive and satisfactory building material. It is common practice in some rural areas to use cow dung, or a mixture of chopped straw and clay, separately or together, as alternative (but not stable) binders. Although these are often satisfactory in a continuously dry climate, all their binding properties can be lost in wet conditions – with disastrous results. These and other organic materials are prone to being washed out or dissolving in water.

4.2.3 Soil composition for stabilization with lime

In this context, 'stabilization' means achieving a mix that will prevent a soil from dissolving under water and will increase its compressive strength.

Many soils containing varying amounts of clay can be stabilized by the addition of lime, because of the way in which the clay reacts with the lime.

Well-graded sharp sand is a common and predictable aggregate often specified for use with non-hydraulic lime for mortars, plasters, and renders. One of the disadvantages of using a mix of lime and sand only, however, is that far higher proportions of lime are required.

A major disadvantage of a sand and non-hydraulic lime only mix is that in wet or flood-risk areas it will not set under water or in very wet conditions (although the addition of a pozzolan to a lime–sand mix makes a hydraulic set possible). Without the addition of a pozzolan, most clean sands will not assist a hydraulic set, although non-hydraulic lime and sand only mixes can produce excellent (un-stabilized) mortars and plasters.

Very silty soils with little or no clay or sand content will seldom be a suitable building material and should be avoided. However, there is a wide range of clay-content soils which are suitable for lime stabilization.

4.2.4 Obtaining soil samples

Un-stabilized clay soil mixes, although strong when dry, can turn to soft mud within half an hour or less of being submerged in water, whereas generally, only a small amount of lime is required to stabilize them. Sands used in all stabilized mixes work best if the sand is well-graded and sharp, which is not always available – unless sourced from different strata or location and mixed (eg a sharp hill sand, with a softer river sand).

In areas where there is a tradition of building with earth, enquire locally where the best quality clay soil for building is to be found. It may be that some sources are currently in use and the areas and depths from which good soils can be obtained are already known. In some regions, particular clays and soils are selected for decorative work and plaster, whereas other clay soils may be used specifically for brick- or block-making. Local knowledge is invaluable.

There are some natural indicators of clay soil:

- If there is a pond or lake nearby, their water-holding capabilities suggest clay is present.
- If, after rain, puddles formed on the soil surface remain much longer in particular areas, soil from these areas is likely to have a clay content.
- If the soil cracks and curls up when it dries out, this indicates the presence of clay.

Before a mix is designed, the soil will need to be tested to establish its composition, particularly its clay content, and whether it contains anything that will prevent long-term set and stabilization.

A representative sample of soil may be obtained by digging a trial hole. This should ideally be undertaken in areas that have been identified as suitable locations for larger-scale soil extraction related to the main work. All topsoil and organic matter from the trial area should be removed first. Dig deeper to inspect the lower layers of soil where there is likely to be a colour change and where the subsoil has no smell (see Figure 4.33).

If there is extensive rock and large gravel, or the soil is unsuitable in any way, investigate the ground elsewhere. To save work, it may be best to dig trial holes in conjunction with other excavations, such as those for the lime-slaking pit or drainage channels.

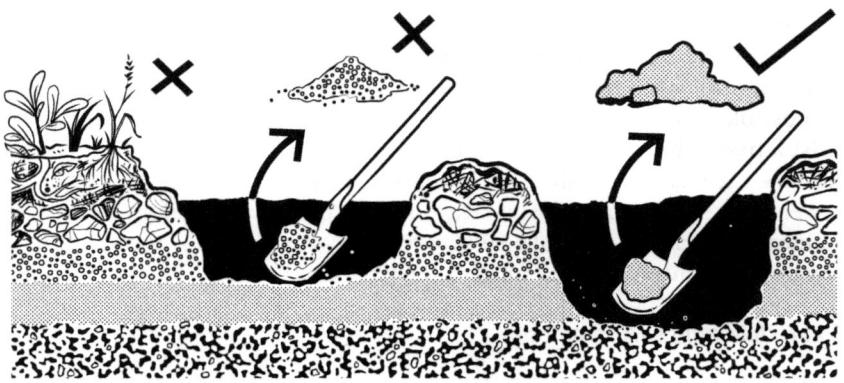

Figure 4.33 Trial holes and soil selection.

Trial holes for soil investigation are best dug with a smooth exposed face facing the sun (south in the northern hemisphere). The different strata can then be inspected in good light as the depth of the trial hole is increased. Test the subsoil from each stratum.

As clay-rich soils are particularly effective for stabilizing with lime, check if there is a brickworks in the vicinity and request a sample of their subsoil for testing.

The following simple field tests can be used to determine soil type and clay content. It is best if the soil is taken from a sufficiently large and consistent stratum to provide an adequate quantity for the main work. Carefully label each soil sample showing location, depth, and date.

4.2.5 Simple field tests for clay content

It is advisable to use several methods of field testing soil, as one test on its own will not necessarily establish whether a soil has a clay content or indicate which soil, dug from which depth, has a greater clay content. Document and label each soil sample with permanent marker, showing location, depth and date.

Before conducting preliminary clay content field tests from a typical sample, sieve out any gravel or stones over 2 mm (ASTM No. 10 sieve), and prepare the subsoil by making it 'plastic', i.e. to plastering consistency. To do this, temper the soil (wet and mix it, so it is damp but not liquid), and ideally keep it damp for between half a day and two days to allow time for the clay to react with the water and other particles in the soil. Turn the damp soil sample over, mixing it well to bring it to an even colour and plastic consistency (a little firmer than that needed for plastering). It is then ready for testing.

Wash test. Rub a sample of prepared damp subsoil between your fingers (see Figure 4.34).

- If you can feel the grains of the soil easily, this indicates a sandy soil.
- If it feels sticky, but it is easy to rub your hands (and forearms) clean when it has dried on the skin and the residue is a fine powder, this suggests a silty soil.
- If it feels sticky and fine, and water is needed to clean your hands when it has dried on your skin, then it is likely to be a clayey soil.

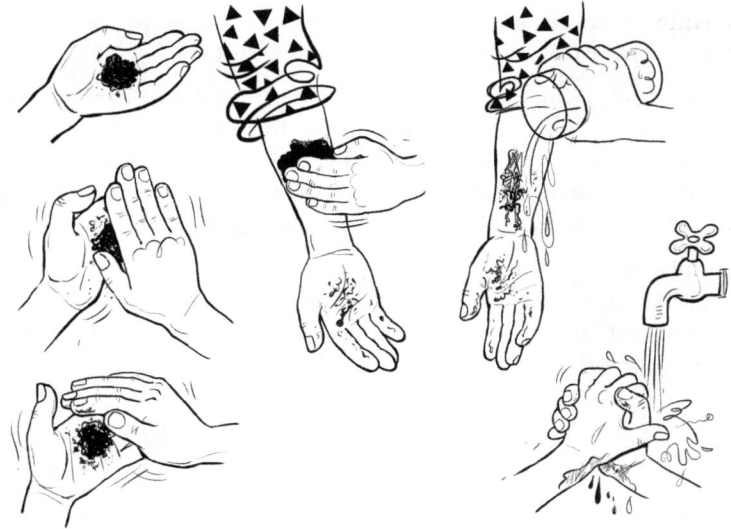

Figure 4.34 Wash test.

Shine test. Form a handful of prepared damp subsoil into a firm fist-sized ball and cut across the ball with a sharp, clean, dry knife. If the ball's cut surface is shiny, the mixture has a clay content. If the cut surface is dull, it has a lower clay and higher sand or silt content.

Additionally, a knife will meet resistance when cutting into a clay soil – it will be noticeably harder to cut through a ball of clay than through a ball of silty or sandy soil (see Figure 4.35).

Granularity test. When rubbed between the fingers, clay feels soapy and silt feels floury. Sandy soil will feel gritty or granular, and will break down quickly. With a little experience, an initial granularity test can be as simple as grinding a little of the subsoil between the teeth or fingers and feeling for grain size.

Figure 4.35 Shine test.

Smell test. Moist clay subsoils have no smell. If a moist sample of subsoil smells damp or loamy (earthy), it is likely to indicate the presence of organic matter. Do not use this in building. Try digging deeper.

Ball-drop test. Form a handful of prepared subsoil into a compressed ball about 50 mm (2″) in diameter. The mixture to be tested has to be as dry as possible, yet damp enough to hold a ball shape. What happens when this ball is dropped from about shoulder height onto a hard, flat surface will give an approximate indication of the soil content (see Figure 4.36).

- If the ball shatters into many pieces, the soil has a very low clay content, and its binding force is poor. It cannot be used as a building material on its own, although it can be modified and its composition adjusted following further tests.

Figure 4.36 Ball-drop test.

- If the ball develops a big crack and possibly some smaller ones, like those of the third ball in Figure 4.36, and stays more or less intact, the soil may have low binding force but may have enough clay and sand content to work well with lime in making renders and earth blocks.
- If the ball is flattened only slightly and shows no cracks, it has a high clay content. It may need the addition of some sand to improve strength. The mix will need a higher proportion of lime for effective stabilization and will likely need fibres to reduce cracking if used for plasters and renders.

Cigar test. Select a representative sample of dry soil from one stratum and eliminate particles larger than 2 mm (ASTM No. 10 sieve) by sieving the soil. Add just sufficient water to make it plastic (as described above) and form a ball of approximately 60 mm or more (about 2″ to 3″) diameter. Roll it into a cigar shape about 20 mm to 25 mm (¾″ to 1″) thick and about 300 mm long (about 1′). Place the cigar-shaped sample on a flat surface (table top) at right angles to the edge, and slowly push it forwards and over the edge. Measure the overhanging length just before it breaks away, or catch it before it falls to the ground and then measure the broken length (see Figure 4.37). If the broken length is less than 50 mm to 75 mm (2″ to 3″), the soil is sandy and/or silty with a low clay content. If it is 100 mm to 150 mm (4″ to 6″), it has a medium clay content; and a break at over 150 mm (6″) indicates a high clay content.

Figure 4.37 Cigar test.

78 BUILDING WITH LIME-STABILIZED SOIL

Ribbon test. A test described by John Norton (1997) uses a similar sample flattened into a ribbon only 3 mm to 6 mm thick. The ribbon is pushed out between and in front of the thumb and forefinger over a flat surface until it breaks. He gives the lengths of continuous ribbon pushed over the edge before breaking as an indication of clay content as follow:

- 300–150 mm (12"–6"): high clay content
- 150–100 mm (6"–4"): medium clay content
- 50 mm (2"): little or no clay content.

The ribbon test, however, is often difficult to perform satisfactorily in the field, especially for those new to working with clayey soil, and it is therefore recommended that the other soil tests described here are conducted first.

Disc test. This is a relatively quick field test which gives an indication of approximate clay content by comparing differential shrinkage and the resistance of soils when they are dried (see Figure 4.38). To carry out the test:

1. Select subsoil samples for different depths or strata.
2. Crush or remove from the sample any lumps and particles larger than 2 mm by sieving (ASTM No. 10 sieve) and prepare the soil as described above.

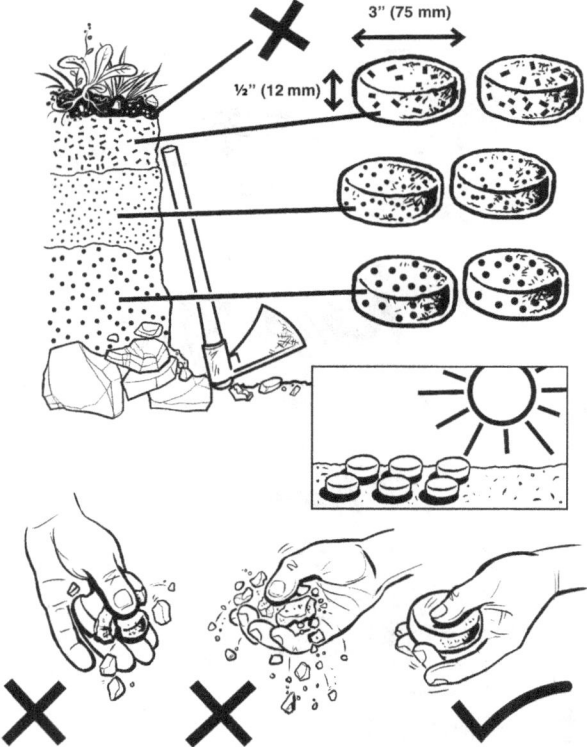

Figure 4.38 Disc test.

3. Make small balls, about 40–50 mm in diameter (2"), and form into flat discs inside 10–15 mm deep (¼" to ½") plastic sleeves. These could be cut from a half-litre (1 pint) plastic water bottle.
4. When the discs have completely dried out, compare the percentage by which each has shrunk, and assess their resistance by trying to crush them between finger and thumb:
    - If there is no shrinkage and the disc crushes easily to powder, the soil is sandy.
    - If there is some degree of shrinkage and the disc crushes to powder, the soil is silty.
    - If there is noticeable shrinkage and it is difficult to crush the disc, the soil is clayey.

Sedimentation test (also known as the 'jar test'). This is a useful test giving a visual guide to the approximate proportions of different constituents in the soil (see Figure 4.39).

1. Fill a cup with relatively dry subsoil from one recognizable soil stratum. Pick out any large gravel and stones. Crush any lumps with a hammer or a length of wood until the soil is all crushed to the same size. Sieve the soil through a 0.45 mm (ASTM No. 40) or other small sieve (no greater than 2 mm or ASTM No. 10).

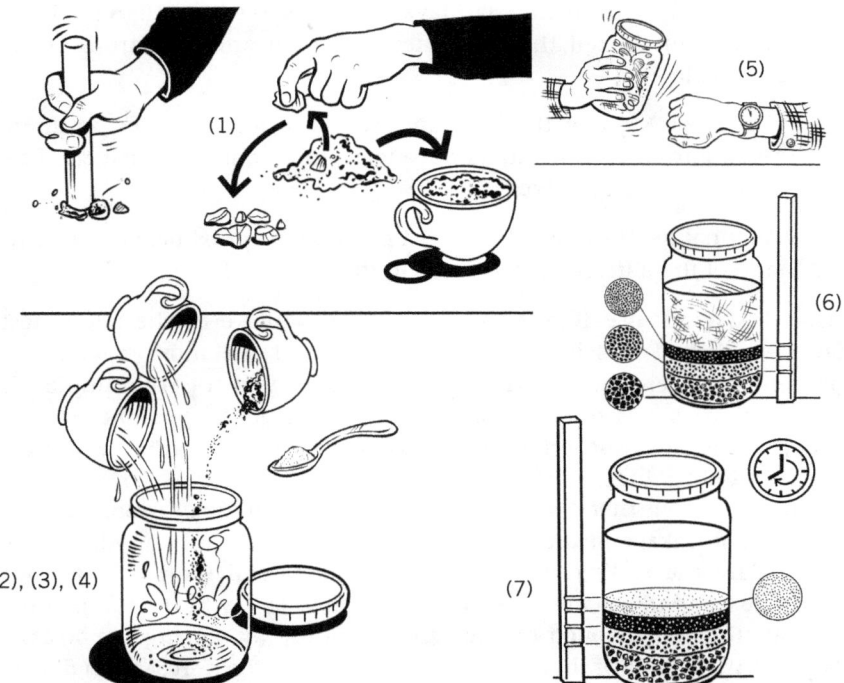

Figure 4.39 Sedimentation test (jar test).

2. Place the subsoil in a tall transparent and lidded jar until the jar is about one-third full. The jar should have straight sides and a flat base to assist an accurate reading of proportions.
3. Add water until the jar is three-quarters full.
4. Add half a teaspoon of salt (this will help the microscopic clay particles settle out of the water quickly).
5. Shake the jar hard for two minutes to separate all the particles.
6. Allow the material to settle out – there will be initial settling out followed by a longer time (maybe 12–24 hours or more) for the water to clear as the particles settle. It is important that the jar is not moved or disturbed at all while the contents settle.
7. Observation of layers:

   Sand layers: Larger sand particles will fall to the bottom almost immediately and the finer sand particles will follow. The variation in particle size can be observed by eye. Sand with different-sized grains makes a better, stronger material.

   Silt layer: After about 20 to 30 minutes, any silt is likely to have settled on top of the sand, making a separate layer with a slightly different colour to that of the sand. At this point, carefully mark the top of this layer with a permanent-marker line on the side of the jar, taking care not to disturb anything. This will help distinguish the silt layer from the clay layer that will settle out later – it can be difficult to tell the difference if the silt and clay are both the same colour.

   Clay layer: Most of the very fine clay particles will often settle out overnight (after 8–12 hours or longer), more quickly if salt has been added as described above.

   The final depths of each layer gives a visual, approximate guide to the ratio of the different constituents in the soil sample.

Linear shrinkage box test. When one or more of the above tests have established which soils contain clay, and the most suitable soil (or soils) has been selected, the percentage of clay in the soil sample (the clay fraction) needs to be established as accurately as possible in order to determine the appropriate lime proportions for the initial trial mixes (Stage 2 field testing).

The next step is therefore a field test to measure to the nearest millimetre the linear shrinkage of the most suitable soil(s) when fully dried, in order to calculate the soil's clay fraction. A prepared soil sample in a linear shrinkage box will usually take about seven days to dry in the sun in a hot, tropical climate (although drying can be sped up by placing the box in a warm oven). Details of this important field test are given in Section 5.3 and Figure 5.4.

FIELD TESTING: STAGE 1 – MATERIALS   81

4.3 Sand

4.3.1 Simple field tests for sand

An initial field assessment of sand can be carried out without any tools other than a few accurately graded sieves; these will also be valuable for the main work. Appropriate sieve sizes for grading particular materials are set out in Table 4.1.

Inspection and observation field test. If clean sand is rubbed between moist hands it should leave no stain. The size and sharpness of grains can initially be judged by eye and feel. The ideal sand has an optimum mixture of grain sizes from coarse to fine (see Figure 4.40). Wide mortar joints and lime concrete, for example, require a proportion of grit or gravel with a grain size in the order of 3 mm to 5 mm (ASTM sieve sizes No. 6 to No. 4) or more. Generally, the fineness of the sand and aggregate selected should relate directly to the fineness or thickness of the work and the finish required.

Nearly all sands will be composed of grains with a mixture of different sizes, but, for the purpose of judging sand by eye, coarse sand particles may generally be assumed to be between 5 mm and 2 mm ($^3/_{16}''$ and $^3/_{32}''$), and medium sand between 2 mm and 0.65 mm ($^3/_{32}''$ and $^1/_{32}''$). Very fine sand or dust will have particles that range from 0.65 mm ($^1/_{32}''$) down to 0.06 mm

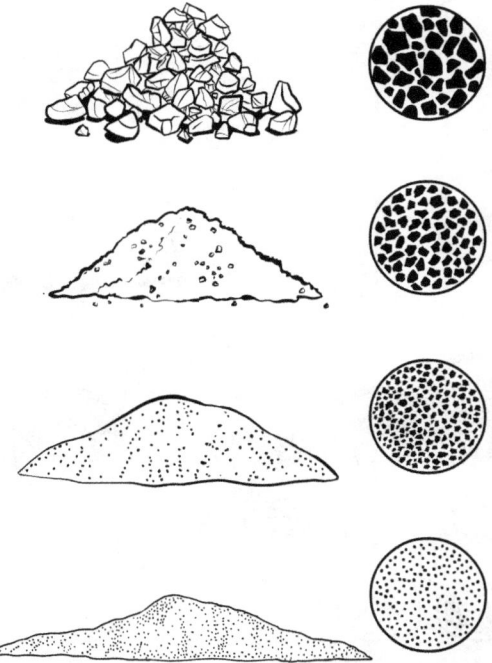

Figure 4.40 Particle-size grading. An aggregate with well-distributed particle sizes will help to strengthen a mix.

82 BUILDING WITH LIME-STABILIZED SOIL

(60 microns). Material containing even smaller particles falls into the category of silt or clay, the proportion of which can be approximately determined by the sedimentation test described above in Section 4.2.5.

Sand particle size analysis. Accurate sieves are required for this simple test to check that a sand has the recommended particle-size distribution for the best quality work. A series of up to eight sieves is required to field test all materials. Research laboratories carrying out particle-size analysis will be equipped with a series of different-sized sieves stacked and operated mechanically – more sieves than are generally used for manual testing. However, the number of sieves for field testing sand can be reduced only to possibly three or four, each of a diameter about 250 mm (10″). The weight and size of equipment can be further reduced by using interchangeable sieve mesh bases for a single frame.

The amount of sand passing through a 5 mm (ASTM No. 4) sieve but retained on and/or passing through 2 mm (No. 10), 0.6 mm (ASTM No. 30),

Figure 4.41 Particle-size analysis of a well-graded sand.

Table 4.3 Particle-size distribution of well-graded sand.

Sieve size/mm	% of sand passing sieve	Well-graded sand adjustment
6.30	100	Higher proportions of coarse sand are suitable for rammed earth, cob, blocks, and render base coats.
5.00	95–100	
2.36	60–90	
1.18	30–80	
0.6	15–70	Higher proportions of finer sand are suitable for render finishing coats, internal plaster, bricks and mortar.
0.3	5–50	
0.15	0–15	
0.075	0–5	

and 0.106 mm (ASTM No. 140) sieves will give an initial indication of the overall particle-size distribution, i.e. the relative quantities of coarse, medium, and fine sand in the sample (see Figure 4.41 and Table 4.1). British Standard 1377-2:1990 gives sand particle sizes as coarse 2 mm to 0.6 mm, medium 0.6 mm to 0.2 mm, and fine 0.2 mm to 0.06 mm.

An example of well-graded sand is given in Table 4.3.

4.4 Pozzolans

4.4.1 Reactive minerals

In order to achieve the stabilization of a soil, sand or other aggregate with little or no clay content, a material that is reactive with lime can be used. Reactive materials of this nature are termed pozzolans.

Natural and artificial pozzolans occur widely and are rich in reactive minerals, predominantly silica and alumina. Natural pozzolans can be the result of volcanic action. Artificial pozzolans can be produced by grinding and/or sieving burnt waste products to very fine powder or dust. These wastes can be materials such as fly ash, broken clay tiles, or pottery, but caution is needed to ensure these have been well burnt. Usually, the most readily available sources of pozzolan are dust from finely crushed soft-burnt brick, rice-husk ash, pumice, or fuel ash from brick kilns or other local industries.

Brick dust made from a suitable clay that has been well burnt but fired at a low temperature (700–800°C/1,300–1,500°F) is one of the most common sources of reactive pozzolanic material. It is important that the brick dust is from a fully fired brick, and not one that is under-burnt or over-burnt. To prepare a typical pozzolan, waste brick dust is collected from a brickworks and sieved, or damaged bricks or clayware pots are crushed to a powder. Waste fuel ash from various agricultural burning processes is worthy of investigation, such as rice-husk ash, and from industry, although beware of the possibility of pollution and contamination from industrial heavy metals. The ash from a rice-husk-fuelled brick kiln using the clamp firing method was included in lime-stabilized soil blocks in Pakistan in 2014 and 2015. The blocks remained stable under water and gave a hand-held penetrometer wet compressive-strength test result of over 4 N/mm² (600 psi) (Holmes and Rowan, 2015).

84 BUILDING WITH LIME-STABILIZED SOIL

4.4.2 Field tests for pozzolanic reactivity

To stabilize and to achieve a hydraulic set with soils that have a very low clay content, sandy soils or sand, using non-hydraulic lime, the pozzolan needs to be sufficiently reactive. An initial test for pozzolanic reactivity using milk of lime is outlined below. Another method is to carry out an immersion test on a cured non-hydraulic lime and pozzolan sample mix to confirm insolubility and determine wet compressive strength. Lime to pozzolan ratios of 1:1, 1:2 and 1:3 could be tested. Due to the curing time required, this immersion test takes longer than the pozzolanic sedimentation test outlined below but will give a more conclusive result.

Sedimentation field test. This tests the reactivity of the pozzolan with lime to form insoluble compounds, and is illustrated in Figure 4.43. Milk of lime (lime putty thinned to the consistency of milk) is poured into a tall narrow glass or jar until it is one-third full. This is followed by an equal measure of the pozzolan sample, which has been finely ground (to pass through a 0.180 mm (ASTM No. 80) sieve). The finer the pozzolan is ground, the more reactive it is likely to be (Figure 4.42).

Figure 4.42 The finer pozzolanic material is ground, the greater its reactivity is likely to be.

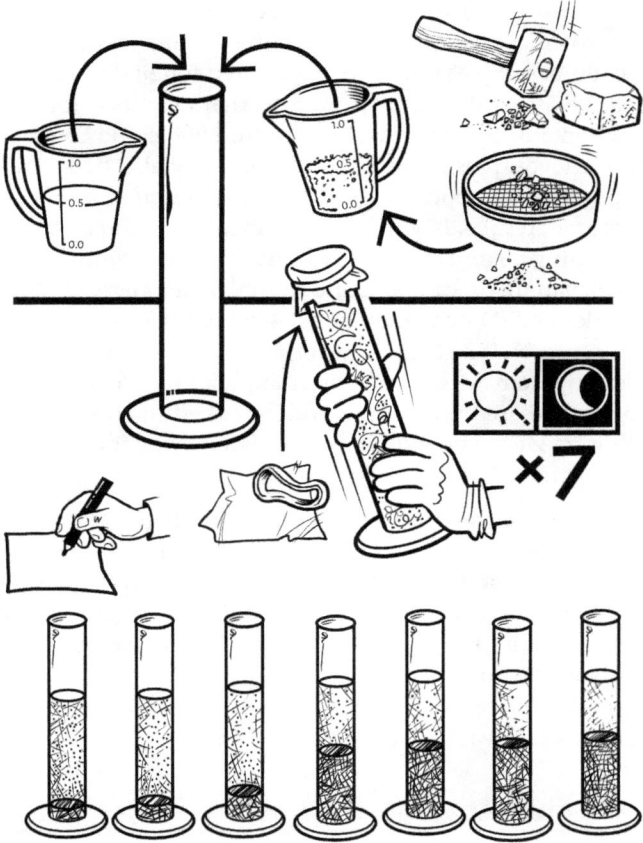

Figure 4.43 Sedimentation test for pozzolanic reactivity.

For a comparative test, it is important that the milk of lime is the same consistency for testing each pozzolan. One way to achieve this is to test all pozzolans at the same time, using a series of similar containers and the same milk of lime mix. If precise measurements are possible, use the following quantities: 0.5 g pozzolan, 0.3 g slaked lime, 200 ml (8 fl. oz) drinking water (Cowper, 1998 [1927]). If this cannot be done, ensure that the lime has the same reactivity and the milk of lime has the same specific gravity (density) for each of the comparative tests.

Seal the end of the container and shake for two minutes every 12 hours (i.e. in the morning and last thing in the evening) for one week. Measure the depth and observe the bulk of the sediment shortly after shaking. Compare this with a fresh mixture of the same material or with another pozzolan given the same treatment. After seven days, the increase in the volume of the solid matter, which can be measured by the change in height of the sediment column in the jar, will indicate the extent of pozzolanic reactivity.

Immersion and compressive-strength tests. A more accurate method of testing pozzolanic reactivity is by comparing the compressive strength of lime and pozzolan mix samples that have been cured for 28 days. The compressive strength can be assessed in similar ways to that described for testing mortar mixes in Section 5.8, and a hand-held penetrometer could be used to measure the wet and dry compressive strength. Cubes may also be immersion tested (Section 5.5 details methods of preparing 50 mm cube moulds). To obtain the most accurate results, laboratory equipment is required. A laboratory test involving crushing 50 mm (2″) cubes to determine the compressive strength of lime and pozzolan mixes can be found in Indian Standards IS 1727, 1344 and 4098 and Nederland (Dutch) Standard N 488 (1932).

As an example, N 488 gives a test for pozzolanic reactivity based on rapid strength development with trass (volcanic tuff) as a pozzolan. The field test outlined below is based on the method described in this standard, extracts from which were kindly provided and translated by Michael Wingate. For full details, see Appendix 5.

- Test pieces are made up in the proportions of two parts by weight of trass (pozzolan), one part by weight of hydrated lime, three parts by weight of standard sand and from 0.90 to 0.95 parts by weight of water.
- The test pieces are then cured and hardened at a controlled temperature of 15°C (variations from this temperature would have a marked effect on the strengths achieved).
- The minimum tensile and compressive strengths of the test pieces should be as shown in Table 4.4.

Table 4.4 Minimum strengths for N 488 test samples.

	Tensile strength kg/cm^2	Compressive strength kg/cm^2
3 (days) in the air and then 11 (days) under water	12 (1.2 N/mm^2)	50 (5 N/mm^2)
3 (days) in the air and then 25 (days) under water	16 (1.6 N/mm^2)	70 (7 N/mm^2)

The standard also suggests making a preliminary assessment using a Vicat (needle) test. Again, full details are given in Appendix 5.

Field testing for pozzolanic reactivity is a practical application of this standard. As such, possibly with minor modifications, it is suitable in warm and tropical climates. Where comparative testing is carried out in cold climates, an allowance for heating will probably be necessary to maintain the temperatures required.

4.5 Fibres

Several building components benefit from the addition of fibres in their mixes, particularly render, as fibres increase the tensile strength of the finished material. Chopped straw is typically used in this way although there are harder-wearing fibres which are likely to prove more durable in wet conditions such as sisal, jute, hemp, and hair. Rice straw and dried reed (both of which grow in wet conditions) or chopped grass may be a suitable alternative, subject to testing. Jute and hair have a long history of effective use in lime renders and plasters, and straw in earth plasters and daubs. However, short chopped straw has been found in well-preserved lime plaster in Kot Diji Fort, Khairpur, Southern Pakistan, constructed over 200 years ago.

The length and size of fibres need to relate to the size of the building element for which they are being used. For plaster backing coats, fibrous material is cut into lengths of about 25 mm to 50 mm (1" to 2") with a knife or a chaff-cutting machine (Figure 4.44). The chopped fibres help prevent plaster and other elements cracking when they dry. (See Section 6.8 on making trial render and plaster panels for testing mixes with different proportions of fibre.)

Fibres should be evenly spread throughout the mix as it is turned over, to ensure they do not clump together, which would create weaker areas within the plaster.

Figure 4.44 Producing chopped fibres.

4.6 Additional materials

Historic references to the addition of organic material – including egg, blood, fats, oils and cow dung – for improving mixes and finishes are common. The tendency of clay-content soils to shrink and crack when they dry, sometimes with a corresponding detachment (falling off) of render from the

88 BUILDING WITH LIME-STABILIZED SOIL

wall, can be reduced by the addition of some organic materials, although care must be taken to avoid excessive proportions of these and reducing vapour permeability, which the addition of oils or fats would cause.

4.6.1 Cow dung

The use of cow dung for soil modification is widespread. It is currently regularly used in many rural areas of the world in conjunction with chopped fibres or on its own to reduce shrinkage and improve the initial tensile strength, workability, and adhesion of earth bricks and blocks, cob walling, and render. Cow dung was, and is, frequently used with lime and sand or marl for lining around and inside fireplaces, flues, and other areas close to and subject to heat (parging). Archive records indicate that proportions of lime to cow dung for parging chimney flues varied as much as between 1:3 and 3:1, sometimes with the addition of sand.

The excellent binding and adhesive properties of lime and cow dung were used widely in England up to the latter half of the 20th century, and their use continues in conservation. Cow dung acts as a binder and improves the plasticity of a mix. When used in conjunction with lime and soil, there is an additional stabilizing effect and a noticeable improvement in weather resistance. The significant constituent of the dung is a mucus which reacts with lime to form a gel. The gel both stabilizes the clay mineral wafers and supports the lime and sand until the carbonation and stabilization processes are complete and the final strength obtained. Historic references suggest that cow dung and probably other forms of animal dung also have a degree of heat resistance when combined with lime.

Where the addition of cow dung is an accepted and appropriate practice, it is therefore included in some of the proposed trial mixes for testing. Its

Figure 4.45 Fibres and organic waste products may be included in trial mixes, particularly for renders and where local tradition has shown this to be of benefit.

addition may be especially useful in external renders and roof finishes, improving wet-weather resistance and helping to minimize cracks. These are important design considerations for roof screeds over a large surface area on a less than solid substrate. Although this method is widely used in some countries, it is not recommended for roofs; the use of alternative finishes, such as thatch or tiles, is advisable.

Other mixes suitable for testing include higher proportions of cow dung and lime putty with less aggregate. They will need to include clayey soil, pozzolan, or hydraulic lime if they are to be tested for wet conditions.

As an example, Shawn Kholucy (2013), working on Thorpe Hall, Suffolk, found that the most satisfactory of four trial mixes of local materials for chimney parging was one part slurried cow dung and one part lime putty to three parts haired aggregate. His report on the project concludes with an explanation of the chemical reaction obtained by the use of cow dung and lime in the mix, which in the report is expressed by Dr James Yates as:

$$Ca(OH)_2 + K_2CO_3 \rightarrow CaCO_3 + 2K(OH)$$

Although such a mix does not produce a hydraulic set (on its own), the resulting chemical behaviour of the cow dung in reaction with the lime – producing insoluble calcium carbonate and soluble potassium hydroxide – demonstrates the potential improvements in application, weather resistance and longevity that can be achieved by adding cow dung to lime mixes.

4.6.2 Oils and fats

Water shedding. Oils, particularly raw linseed oil, have traditionally been used as an additive to limewash in exposed areas to improve external water shedding. Such mixtures are probably more appropriate for areas that are subject to frequent or heavy rainfall, or have extended rainy seasons. If oils are used, a small amount, about 5%, should be added to the final coat of limewash. The more oil that is added, the less well the limewash will adhere to the wall or allow evaporation.

Similar proportions of tallow (melted animal fat) or casein (milk protein) have been used for the same purpose, and to reduce 'dusting' of the surface. Unfortunately, untreated animal-based products may experience mould growth when applied in areas of persistent high humidity or condensation. This risk can be reduced by the addition of a mould inhibitor, adding disinfectants to the mix or by boiling organic substances first.

Long-term maintenance. Adding oil-based materials to the last coat of limewash reduces both the porosity of the wall and the bond for further applications. Therefore, before the wall can be limewashed again, the surface will need to be prepared by stiff brushing down to assist absorption and provide a good key.

4.6.3 Water

One of the benefits of lime is its ability to help clean or purify water: water (say from lakes, rivers, or ponds) can be treated with lime to remove impurities. Water from most sources is likely to be suitable for lime slaking and for making lime-stabilized mixes, particularly if it has been treated with lime. However, the purer the water used, the more consistent and reliable the mixes will be, subject always to confirmation during Stage 2 field testing that the mix, including the water, is satisfactory. If in doubt, have the water chemically analysed in a laboratory, as, in some cases, an excess of salts or other chemicals from, for example, agricultural run-off or industrial waste leaching into the water may be detrimental to set or stabilization. Various field investigation methods and tests to check for this are set out in Section 4.7.1.

Saline water should not be used to slake lime or to make lime-stabilized mixes. In the absence of clean water, saline water may be used to cure finished lime-stabilized building elements (i.e. for damping down), although clean water is preferable.

4.7 Detrimental conditions

4.7.1 Salts and sulphates

Generally, if salts are present in water or the soil they are likely to crystalize on the surface of the soil as a powder. Excess salts in the mix may damage or, in severe cases, destroy the finished work. There is, however, an exception: the adhesion of limewash may be improved by the addition of common (sea) salt in the proportions of approximately 1 part salt to 10 parts limewash by weight (Cowper, 1998 [1927]).

Salts and chemicals that may cause defects include calcium nitrate, calcium sulphate, sodium sulphate, sodium chloride, calcium chloride, sodium hydroxide, nitric acid, ammonia, and other nitrates which may result in expansion of crystals or acid pollution. Laboratory analysis of the soil is recommended if field tests or local knowledge indicate that any of these are present.

If sulphates are present in the soil, they may react with the lime to produce ettringite. This can take in a substantial amount of water and so cause the finished material to expand, although the effect of sulphates on lime-stabilized material is unlikely to be significant for concentrations of total sulphate (expressed as SO_3) of 1% or less. Soluble sulphates in the soil may come from waste industrial products, including agricultural fertilizer, or naturally occurring minerals, most commonly gypsum. Gypsum may occur in small, localized deposits and its solubility and concentration may vary over short distances.

Research in the USA (ARBA, 2004) has confirmed that sulphate-induced heave is the cause of some failures. The research concluded that lime could be expected to be a satisfactory stabilizer not only for most clay-bearing soils, but also for those with sulphate concentrations up to 7,000 ppm (parts per million). Soils with sulphate concentrations of up to 20,000 ppm can also be stabilized with lime if modified by the addition of ground granulated blast-furnace slag (GGBS).

In some areas, geological mapping identifies the location of sulphate-bearing soils as well as the many regions that are more suitable for lime stabilization. Where this information is not available, further field-testing options need to be considered, starting with observation and initial swell testing of soil samples. Such field observations and tests for the presence of salts, particularly sulphates, have been used in connection with the construction of road sub-bases and other civil-engineering projects. Some of these methods are appropriate for alternative uses of lime-stabilized soil, as outlined in Chapter 7.

4.7.2 Testing for the presence of salts

There are a number of tests to determine whether salts are present. These range from simple field tests and observation to those requiring laboratory conditions and analysis. Initial field tests to give a general indication of the type of salts present can be carried out with a variety of specialist equipment and reagents. For a more detailed analysis, and to establish concentration levels, controlled laboratory conditions are required. The results of Stage 2 field testing of mixes, and of observation after curing, immersing, and drying samples, should indicate whether a detrimental amount of salt is present. Cube samples should therefore be examined for salt efflorescence as well as expansion, loss of strength, and cracking, on completion of the test procedures described in Chapters 5 and 6.

Detrimental materials may be neutralized or washed out of some soils. However, doing this may not always be cost-effective, and laboratory analysis may be required to establish whether successful modification of the soil is possible. Alternative soil sources should be considered where this is the case.

Field observation test. Gypsum (calcium sulphate) can appear in the soil as visible crystal formations, incorporating crystals of various sizes up to 5 mm (¼") or more. It may also take the form of microscopic grains which can be seen as efflorescence or a white powder on the surface, particularly in dry conditions.

Swell test. Methods of testing for volumetric expansion by determining vertical swell and swell pressure are set out in detail in ASTM D 4546-96. Simple field tests, whilst not as accurate, will indicate, firstly, whether expansive salts are present and, secondly, if they are sufficiently damaging to require soil modification. If this is the case, the use of alternative soil sources should be considered. This principle can be applied initially to sample cubes and blocks of lime-stabilized soil, prepared as described in Chapter 5. The extent of swell, shrinkage, or breakdown (if any) can be measured during and after the 28-day curing period, and a decision made on whether any form of modification is necessary.

As a preliminary field test, specifically to determine the extent of swell and swell pressure, prepare a minimum of three test specimens, and preferably six, for 28-day curing. First pass the soil through a 0.45 mm sieve (ASTM No. 40) and make up the lime-stabilized soil mix. The moisture content should be optimum for two mix specimens, slightly below for two samples and slightly above for the other two (the ball-drop test described in Section 6.3.1 can be used to check this). Compact the mix into strong moulds 150 mm (6″) high and either 150 mm (6″) in diameter or square. National standards for compaction with a hand rammer recommend that the test samples are built up in three layers and that each layer is given approximately 50 blows with a compaction hammer. The samples should then be measured for vertical expansion, taking readings immediately after full immersion, every minute for half an hour, and then throughout the curing period with the interval between readings gradually increasing to three days. A final reading should also be taken at the end of the 28 days.

This test can be extended by loading the specimen with weights to produce a pressure in N/mm^2 (or psi) equivalent to the compressive strength of the building element for which the mix is required. Any swell following immersion, or deformation on loading, should be recorded and related to that of soils with known sulphate content. The USA's National Lime Association (NLA) considers three-dimensional expansion of below 2% after lime treatment to be acceptable for expansive soils.

Testing with reagents and indicators. The availability of reagents in the field may be limited depending on outside sources and an understanding of application methods. Probably the most convenient are paper indicator strips, such as those for assessing the acidity of a soil: when dipped in a soil solution, these strips indicate a soil's pH by colour change alone.

A reagent used to determine the level of sulphates present is barium chloride. Using this involves diluting the soil with distilled or deionized water at a ratio of 1:10, boiling for 24 hours, filtering, adding barium chloride, further heating and filtering, and then collecting, washing, and weighing the precipitated barium sulphate. This form of analysis is more suited to the laboratory than the field.

Conductivity test method. For the conductivity method, the field detection of sulphates in soil relies on a portable conductivity meter powered by a nine-volt battery. The Texas Department of Transportation brought such a meter into official use in 2005: detailed preparation and calibration procedures for the conductivity meter, following the manufacturer's instructions, are set out in TxDOT Designation: Tex-146-E. Soil samples are obtained and prepared in the standard way, including sieving through a 0.452 mm (ASTM No. 50) sieve. The 4 oz (113 g) soil sample is mixed with 3.5 oz (100 ml) of distilled or deionized water and shaken for one minute. Meter readings are taken and the conductivity of the water subtracted from that of the soil-water solution. The test is repeated after 12 hours.

Soluble salts testing. The soil sample is thoroughly dried, preferably in an oven, and then weighed. It is put into a container that includes a very fine filter which will allow water to flow through the soil without any of the sample being washed away. Hot (80°C) water is gently passed through a conveniently sized sample of the soil, taking care that no silt or clay is lost. The soil sample is then dried fully and re-weighed. The weight of soluble salts is the difference between the two weights and may be expressed as a percentage of the weight of the initial soil sample.

4.7.3 Acidity

The addition of lime to a soil modifies it in several ways, one of which is to reduce its acidity. The more acid a soil, the more lime will be required to stabilize it. ASTM standard D 6276 sets out the Eades and Grim test which advises that the amount of lime required for stabilization should be approximately that which increases the pH of the treated soil to at least 12.4 (see Section 5.4.4).

4.7.4 Clay swell

Materials that lead to instability or a weakness of the finished building element need to be identified and eliminated from mixes; clays with excessive swell may be such a material. There are many types of clay minerals however, many of which are considered to be without swell, or with a level of swell that can be eliminated or significantly reduced by modification.

> *For example, those with little or no swell include kaolinite, chlorite, and allophane. Those that may produce extensive swell are smectites and montmorillonites. There may also be a degree of swell with some of the illites, muscovites, and vermiculites.* (Beetham et al., 2015)

In many cases the swelling properties of untreated natural clay may be offset or eliminated by the introduction of lime into a mix, which is particularly effective when used in quicklime form.

Clay swell on submersion and at each stage of curing may be measured in the field using the swell test described in Section 4.7.1 for samples prepared both with and without lime. If the swell occurs in test samples prepared without lime, it is likely to be due to minerals in the clay. Test samples prepared with the addition of lime may also swell for this reason, but the expansion could also be due to the presence of excess salts, possibly sulphates, in the clay mineral.

Field tests of trial mixes with varying proportions of lime (and pozzolan) will establish whether clay swell can be reduced or eliminated. The precise mineralogy of clays and soil salts is complex, but may be determined by laboratory analysis if necessary. Field testing samples as detailed in Chapter 5 can provide basic information on clay suitability.

CHAPTER 5
Field testing: Stage 2 – Lime stabilization of soils

5.1 Selection of materials

The completion of Stage 1 field testing should identify a range of satisfactory local materials. Those that are the best quality, and most reactive, can be used for the second stage, which is to design and test mixes that will result in stabilized soil, with or without modification, for a range of building components. To do this, make trial samples of modified soils and appropriate proportions of lime. After curing these trial mixes, carry out a range of field tests to investigate their properties, including insolubility and compressive strength.

First check that the quality of all materials to be used at this stage is consistent and that they meet the following criteria:

- *Lime* – without loss of reactivity and preferably fresh from the kiln.
- *Soil* – having a clay content, sieved for correct particle size for the building element or component, and adequately tempered/matured.
- *Sand* – sharp and well graded.
- *Pozzolan* – reactive with lime.
- *Fibre* – strong, dry and cut to the appropriate length.
- *Water* – clean.
- *Other materials such as cow dung, oil and crushed aggregate* – quality and consistency confirmed and prepared for testing.
- *Salts, including sulphates* – absent, or only present in minimal amounts.

5.1.1 Soil composition

To make satisfactory plasters and building elements, soil should have appropriate proportions of both sand and clay. Sand gives structural strength and clay bonds the ingredients of the soil together. The proportions present in a given soil will generally have been established in Stage 1. If the required proportions do not occur naturally, the soil can be modified by the addition of other materials to the mix.

The addition of the correct amount of lime to the right type of soil can improve its strength and, to varying degrees, its resistance to erosion and stability under water. Lime interacts with recognizable types of soil as follows:

- *Sandy soils* – lime binds the grains of sand together.
- *Clayey soils* – lime reacts with the clay to stabilize the soil.
- *Silty soils* – soils with a very high silt content are not generally suitable for building construction and should be avoided, or be substantially modified by the addition of other soils and/or suitable materials as set out below.

5.1.2 Particle size

The performance and interaction of soils, limes, and pozzolans are, to a large extent, related to their particle size. The precise definition varies from one country to another but, for practical purposes and the field tests described here, they are regarded to be as shown in Table 5.1.

Table 5.1 Indicative particle sizes for soils, lime, and pozzolan.

Material	Particle-size range	ASTM sieve size
Gravel	75 mm to 5 mm	3" to sieve No.4
Sand	5 mm to 0.06 mm	Sieve No. 4 to No. 140
Silt	0.06 mm to 0.002 mm	Sieve No. 230 and below
Clay	0.002 mm and below	Not visible to the naked eye
Quicklime for coarse work (foundations & field tests)	5 mm to dust	Sieve No. 4 to dust
Powdered quicklime	3.35 mm and below (for field test say 5 mm to dust) well burnt (none over- or under-burnt). In practice, compensation for poorly burnt material (preferably avoided) may be by increasing its proportion in the mix.	Sieve No. 6
Lime dry hydrate	0.6 mm and below	Sieve No. 30
Lime putty (and quicklime powder for renders & plasters)	0.180 mm and below	Sieve No. 80
Pozzolan	0.063 mm (63 microns) and below	Sieve No. 230

Lime may need to be finer than above when used for some types of work and finishes detailed in Table 4.1.

FIELD TESTING: STAGE 2 – LIME STABILIZATION OF SOILS 97

5.2 Preparation of materials for trial mixes

5.2.1 Soil preparation

The preparation of materials to investigate their suitability has been set out in Chapter 4. Stage 1 tests can be done with small samples and table-top field analysis. Materials and mixes will need to be prepared on a larger scale for the second stage of testing, which is carried out to determine the optimum proportions of each material in the mix to ensure that the soil is fully stabilized and is appropriate for the building component and purpose intended. The preparation of materials for Stage 2 testing is therefore more comprehensive and detailed than that for preparing samples to test the quality of individual materials.

5.2.2 Soil selection

Inspect different strata down to about 2 m (6'5") or more (see Figure 4.33 in Section 4.2.4 and Figure 5.1 below), and test representative samples of clay-bearing soil from each using the tests described in Section 4.2.5. Select the subsoil these tests have identified as suitable and ensure that there is a sufficient quantity available for the main work, as well as the stabilization trials.

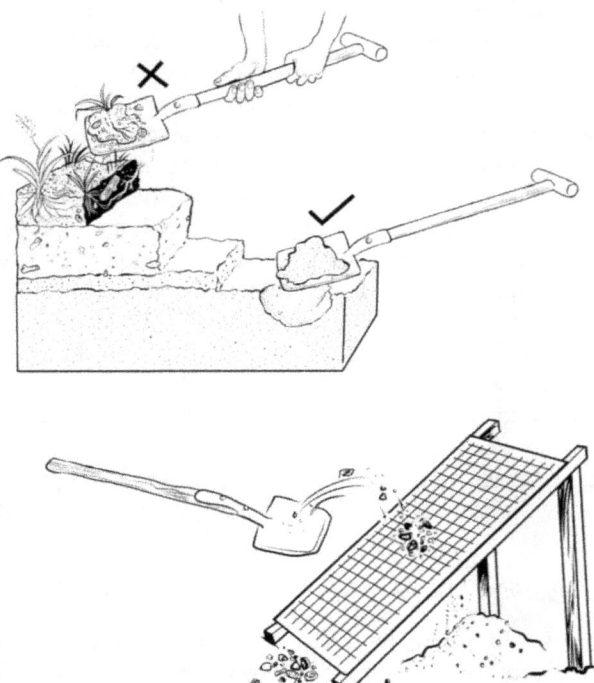

Figure 5.1 Soil selection.

5.2.3 Soil grading

As a starting point for building-element trial mixes, select or modify soil grading as for un-stabilized work. A number of authorities and sources that give soil-grading recommendations in publications on earth construction are listed in the bibliography. Although opinions vary, preferred soil gradings for various building elements fall into the ranges shown in Table 5.2.

Table 5.2 Recommended soil grading for various building elements.

Building elements	Soil component			
	Clay	Silt	Well-graded sand	Fine to medium gravel
Rammed earth	7–15%	15–30%	45–75%	10–20%
Cob	15–25%	10–20%	65–80%	10–15% plus fibre
Blocks	10–30%	10–25%	45–75%	None, consider fibres
Bricks	15–30%	10–30%	40–75%	None
Render and plaster	20–35%	0–10%	60–80%	None, add fibres

Lime stabilization has the effect of widening soil-grading boundaries, particularly when quicklime is used. This is because lime causes clay particles to flocculate and form coarser agglomerates of clay. Following stabilization, the soil will absorb less moisture and compact more readily to give increased strength. This immediate reduction in plasticity and the drying effect, which is most pronounced and accelerated with quicklime, at the same time widens the grading range of soils suitable for use. Soils to be treated with lime may therefore have a higher clay and moisture content (different plastic and liquid limits) than those used for building with un-stabilized earth or other stabilization methods.

To prepare soil for Stage 2 field testing, ensure it is appropriately graded for the building element for which it is to be used (see Table 5.2). Dry the soil and remove all large lumps and stones that may weaken blocks and particularly smaller test specimens. The size of aggregate in the soil should be related to the size of component or building element: that is, very fine for plasters and finishes, much coarser for blocks, and larger still for cob or compacted walling. For testing, break down lumps of soil, which are likely to be clay, into smaller particles. Use a sieve or screen, if available, to ensure that there are no stones or insoluble lumps bigger than 5 mm (¼") in the mixes for larger test specimens and 0.45 mm (ASTM sieve No. 40) for the smaller (50 mm (2") cube) samples.

5.2.4 Soil tempering

For blocks, renders, and plasters, temper the soil before preparing trial mixes. Mellowing for one or two days helps the larger particles to break down, 'mature', and combine well (in some conditions, and for some soils, more time may be beneficial and it can be left longer). Thoroughly mix the soil and leave it to stand overnight. Mix again the next day. If the soil becomes too dry, sprinkle water into it, adding only small amounts at a time so it becomes very damp throughout, but not saturated: it should be slightly sticky and lightly moist. The soil can be left to mature for two days to a week or more provided it is well covered, kept constantly damp and never allowed to dry out (see Figure 5.2).

Figure 5.2 Soil tempering: leave the mix to mature for at least one or two days.

Figure 5.3 Prepare soil mixes in advance; turn the soil and mix again before adding lime.

5.3 Estimation of clay content

In order to obtain stabilization, the proportion of active clay to lime is critical, because of the way lime interacts with the clay minerals. Ideally, for stabilization, the soil should have a minimum clay content of no less than 10%. Many natural soils and clays can be lime-stabilized but it may not be possible to stabilize some others. (Refer to Appendix 4 for the suitability of soils for the addition of lime and Chapter 8 for details of relevant national standards that describe laboratory tests.) Where laboratory testing is not practical, field testing of soils, using methods outlined in Section 4.2.5, to establish whether a soil and/or clay is suitable is essential. Field-testing mixes, as described below, will allow selection of and/or determination of the modifications needed to the soil for its intended use.

To establish lime proportions for stabilization, first determine the exact proportion of clay in the soil, or estimate it as closely as possible. This can be done using a number of different tests. The linear shrinkage test (below) is probably the most practical field test and the best guide for calculating mix proportions.

Whenever possible, the results of these field tests should be verified by laboratory analysis, particularly for repetitive large-scale projects and where there are variations in materials used.

5.3.1 Linear shrinkage test

The linear shrinkage field test is used to indicate a soil's clay content by the extent to which a damp sample (of plastering consistency) shrinks when fully dried.

1. Make a wooden box with internal dimensions of 600 mm × 40 mm × 40 mm (2' × 1½" × 1½"). The box should have smooth internal surfaces, and a base but no top (see Figure 5.4). Lightly oil the inside.
2. Take a sample of soil intended for stabilization, tempered as described in Section 5.2.4, and moisten it to its optimum water content (check this with the ball-drop test described in Section 4.2.5 and shown in Figure 4.36).
3. Use this soil to fill the box. Tamp the soil down, compacting it firmly, and then smooth off the surface (see Figure 5.4).
4. Dry the contents for five to seven days in hot sun, for longer if not hot, or in an oven.
5. When the sample is completely dry, carefully slide all the soil, including any separated pieces, tightly up to one end of the box (see Figure 5.5).
6. Measure the total gap at the end created by the shrinkage in the soil, and calculate the clay percentage as shown in Table 5.3. The total extent of shrinkage indicates the approximate clay content of the soil.

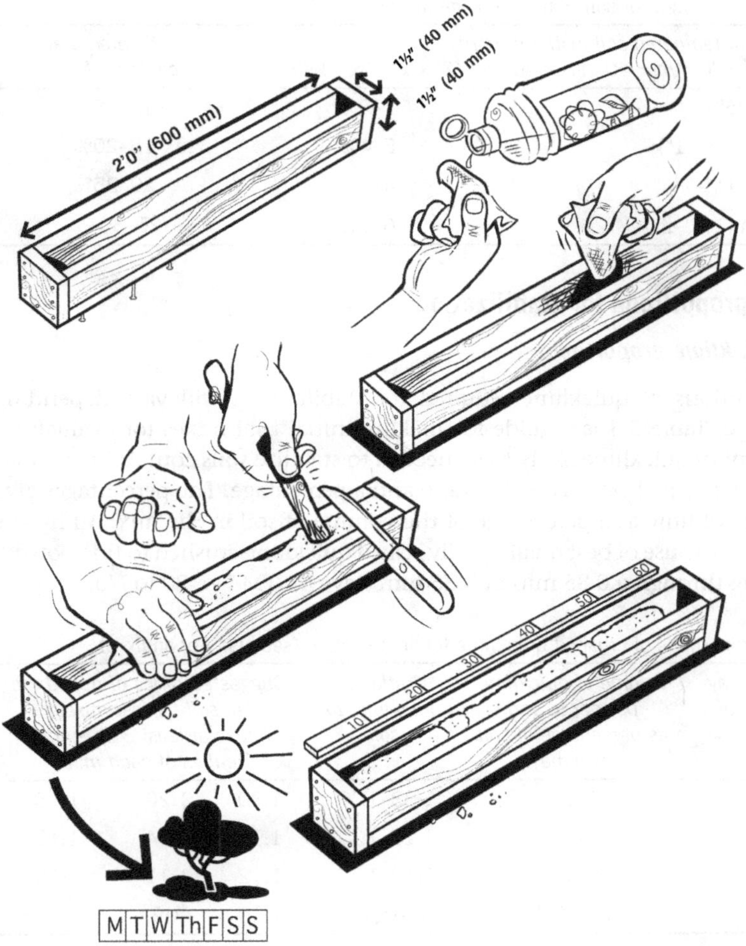

FIELD TESTING: STAGE 2 – LIME STABILIZATION OF SOILS 101

Figure 5.4 Linear shrinkage box test.

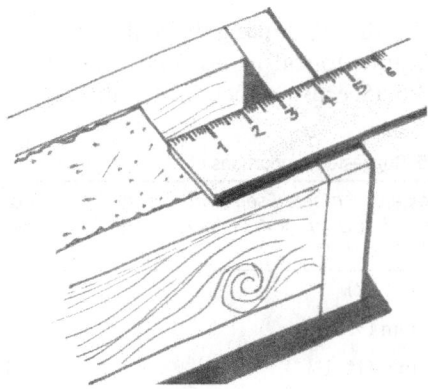

Figure 5.5 Measure the total shrinkage in millimetres when fully dried.

Table 5.3 Clay content of soil from shrinkage measurements.

Shrinkage of tempered soil (without lime) in a mould 600 mm × 40 mm × 40 mm (2' × 1½" × 1½")		Estimated clay content of soil
< 12 mm (½")	1–2%	12–15%
12–23 mm (½"–1")	2–4%	15–20%
24–35 mm (1"–1⅜")	4–6%	20–25%
36–48 mm (1⅜"–2")	6–8%	25–30+%

5.4 Mix proportions for stabilization

5.4.1 Quicklime proportions

The proportions of quicklime required for stabilization will vary depending on soil type. Table 5.4 is a guide for making initial trial mixes for evaluation. The amount of quicklime likely to be needed to stabilize soils containing various amounts of clay is shown as both a ratio and a percentage. The percentages give the volume of lime as a percentage of the volume of soil in the mix. All figures are based on the use of best quality, fully slaked quicklime crushed to fine powder (i.e. passing through a 0.85 mm or 3.35 mm sieve (ASTM No. 20 to No. 6).

Table 5.4 Proportions of crushed quicklime for stabilization (see also Appendix 1).

Estimated clay content of soil	Volume of quicklime powder to add as percentage of soil volume	Ratio of quicklime to soil	Suggested ratios of quicklime to soil for test mixes (make at least 3 specimen cubes of each mix)		
12–15%	3–6%	1:33–1:17	1:30	1:20	1:15
15–20%	6–8%	1:17–1:12	1:15	1:14	1:12
20–25%	8–10%	1:12–1:10	1:12	1:11	1:10
25–30+%	10–12%	1:10–1:8	1:10	1:9	1:8

A table directly linking the millimetre shrinkage to the suggested proportions of quicklime-to-soil to test may be easier to read, and is included below as Table 5.5. (See also Appendix 1, Table A1.2.)

Table 5.5 Suggested proportions of crushed **quicklime** to use in trial mixes (see also Appendix 1).

Shrinkage of tempered soil (without lime) in a mould 600 mm × 40 mm × 40 mm (2' × 1½" × 1½")	Suggested proportions of quicklime to soil for test mixes (make at least 3 specimen cubes of each mix)		
< 12 mm (½")	1:30	1:20	1:15
12–23 mm (½"–1")	1:15	1:14	1:12
24–35 mm (1"–1⅜")	1:12	1:11	1:10
36–48 mm (1⅜"–2")	1:10	1:9	1:8

5.4.2 Alternative trial mixes

An alternative method of calculating optimum proportions of quicklime for soil stabilization tests is to prepare trial mixes where the percentage of powdered quicklime added is consistently 20% (i.e. a fifth) of the clay fraction of the soil.

Table 5.6 Suggested proportion of quicklime for alternative trial mixes.

Estimated clay content of soil	Volume of quicklime to add as percentage of soil volume
15%	3%
20%	4%
30%	6%
40%	8%

As Table 5.6 shows, this results in the use of a slightly smaller quantity of quicklime for the initial trials than the method described in Section 5.4.1.

5.4.3 Lime putty or dry hydrate proportions

For field-testing purposes, the suggested volume of lime putty or dry hydrate in trial mixes is approximately double that of the quicklime that would be used with the same soil, due to the swelling that occurs during hydration. Table 5.7 is a guide for making initial trial mixes for evaluation. (See also Appendix 1, Table A1.3.)

Table 5.7 Suggested proportions of **lime putty or dry hydrate** to use in trial mixes.

Shrinkage of tempered soil (without lime) in a mould 600 mm × 40 mm × 40 mm (2' × 1½" × 1½")	Suggested lime putty or dry hydrate to soil for test mixes (make at least 3 specimen cubes of each mix)		
< 12 mm (½")	1:15	1:10	1:8
12–23 mm (½"–1")	1:8	1:7	1:6
24–35 mm (1"–1⅜")	1:6	1:5.5	1:5
36–48 mm (1⅜"–2")	1:5	1:4.5	1:4

5.4.4 ASTM alkalinity test

ASTM C977-10 sets out a method of establishing the proportion of lime required to stabilize a soil by testing the increase in alkalinity of the treated soil. Lime is added to the soil sample until its pH reaches 12.4, at which point the proportion of lime should be sufficient to achieve stabilization. Colour-changing reagent strips immersed in a weakly buffered solution of the treated soil sample can give an approximate value that should be accurate enough to judge when the pH is between 12 and 13. (See photo 13.)

To obtain a quick reading in the field, dilute a well-mixed stabilized soil sample with fresh water, shake well and, once the soil has settled, insert the test strip for a few minutes. Then read off the pH value. For greater accuracy use a buffered solution of equal parts soil and water, say 50 g of sifted soil to 50 ml of water, (or 2 oz of soil to about 1.5 fluid ounces of water). Shake or agitate it for 5 to 10 minutes, then leave it to stand for about 20 minutes to let the sediment settle. Put the test strip into the water and read off the pH after the colour change.

5.5 Designing the final mix

5.5.1 Finalizing mix proportions

The design of a mix, and particularly the proportion of lime to soil, depends on a combination of soil mineralogy, clay type, particle-size distribution, percentage of clay in the soil, reactivity of the lime, and the volume of water used. A test programme to finalize the best composition for mixes may take time but, in a region where materials of a known and consistent type are available, once the best ratios have been established (and preferably verified by laboratory analysis), they may be used continually.

If an appropriate mix is designed and recorded, it can be reproduced indefinitely for that region. Due to the variation in materials from one region to another, however, it is important to take and maintain thorough records of successful mixes, and the precise source of local materials unique to that region.

Whichever method is used to determine the required proportion of lime for the final mix, it is important that the cured trial mixes are physically tested after curing, as detailed below.

The principal stages in finalizing a mix design are therefore:

- select, prepare and use the best quality materials available;
- calculate the clay proportion for stabilization;
- establish the optimum lime to soil ratio;
- formulate test mixes with soils appropriate for each building element;
- prepare and cure test mixes;
- test the physical properties of cured samples.

Costs can be kept down by selecting soil with a clay content that can be stabilized by the addition of the minimum amount of lime (3% to 10%) or modifying a clay-rich soil (possibly with well-graded sharp sand and/or fibres). All components and building elements incorporating modified mixes must be given a final immersion test, to ensure no further modification is necessary for the finished article to remain stable under water.

5.5.2 Trial-mix samples for each building element

It is recommended that trial mixes for each building element are made in the form of blocks, cubes and discs.

Figure 5.6 Making a block using a trial mix. (See also Figure 6.14.)

- **Blocks** (for brick and block mixes). Dimensions vary according to local custom, but an example standard block size is 290 mm × 140 mm × 90 mm (11½″ × 5½″ × 3½″).

- **Cubes.** For immersion and compressive strength testing, make small 50 mm (2″) cubes to test mixes for mortars, floor screeds, bricks and blocks, and large 150 mm (6″) cubes to test foundation and slab mixes (note that these are in addition to the trial-mix blocks above). Small cubes can be formed in three-gang moulds, made to the above internal dimensions from smooth-surfaced timber and oiled for ease of demoulding. The larger 150 mm (6″) cube moulds produce samples that are a useful size for initial field evaluation and later laboratory testing of foundation and rammed earth or cob trial mixes which contain larger aggregates. Larger 300 mm high × 300 mm diameter (12″ × 12″) cylindrical moulds are also useful for testing these larger aggregate mixes. Further samples of successful trial foundation, brick, block, and mortar mixes should be created using steel cube moulds for laboratory compressive-strength testing and chemical analysis where laboratory facilities are available.

- **Discs.** Render, screed, and plaster trial-mix samples are prepared in the form of discs for immersion and permeability testing, as well as impact testing for floor screeds and wall finishes. These should be about 75 mm (3″) in diameter and 25 mm (1″) thick.

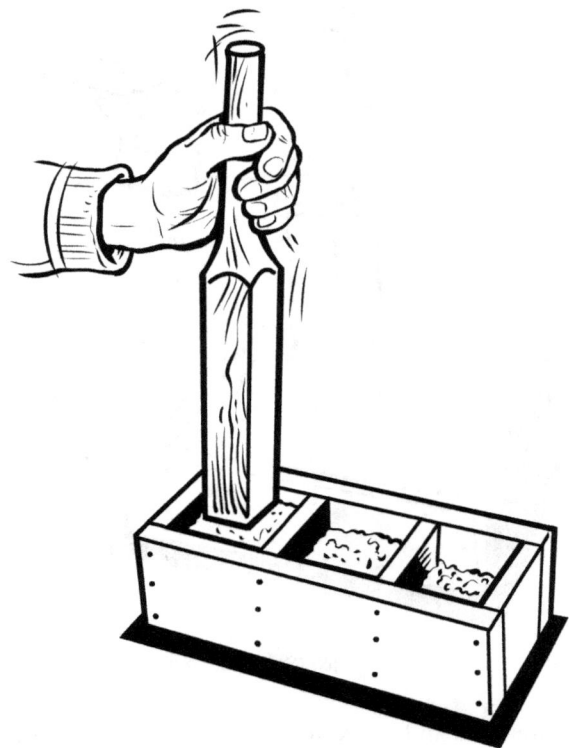

Figure 5.7 Making samples in a three-gang cube mould.

For testing, make up a minimum of three samples of each of at least three trial mixes for each building element. The different mixes for a particular element will have slightly higher and lower proportions of lime, as well as the optimum. This is in order that field testing alone can establish the minimum quantity of lime to stabilize a particular soil. Note that in some cases reducing the quantity of lime in the mix can improve results.

Table 5.5 recommends three different trial mixes based on the linear shrinkage reading, so, for example, for a soil that shrank by 18 mm, the three different mixes would contain:

- 1 part quicklime and 15 parts soil (three samples)
- 1 part quicklime and 14 parts soil (three samples)
- 1 part quicklime and 12 parts soil (three samples).

Of which three samples would be made for each trial mix. A total of at least nine trial-mix samples will therefore be made for each building element. These trial-mix samples should be in a form and size suitable for the building element for which the mix is being tested (generally cubes, discs, and/or blocks, as shown in Figure 5.8). (See photo 17a.)

FIELD TESTING: STAGE 2 – LIME STABILIZATION OF SOILS

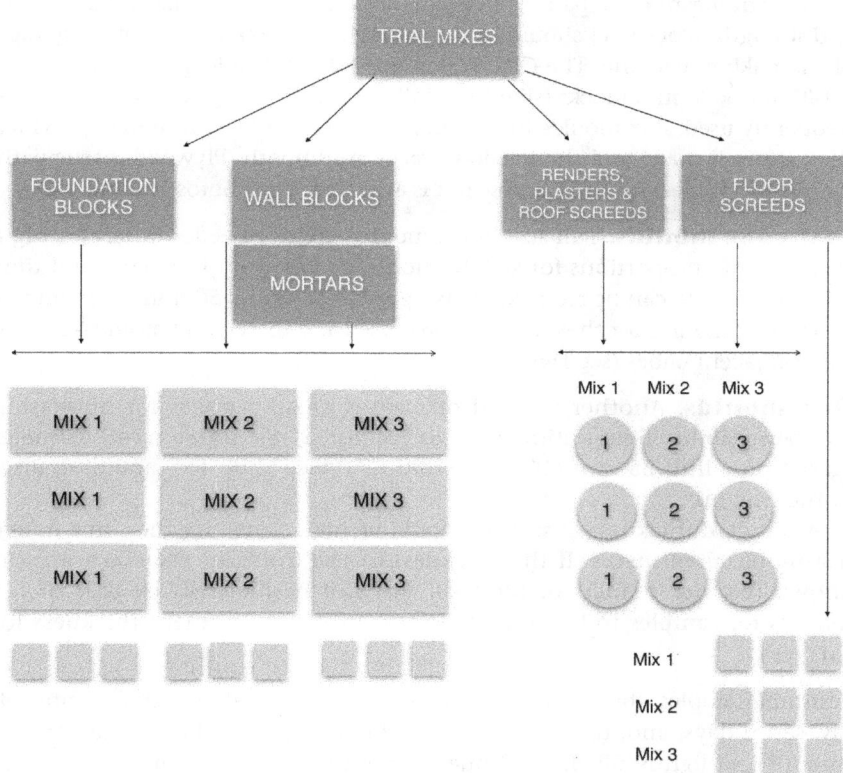

Figure 5.8 Trial-mix samples. Make an initial batch of three samples in the form of blocks, cubes and/or discs of three different mixes as suggested in Tables 5.5 and 5.7.

5.5.3 Sand selection

Select a sand, or modify the soil with an aggregate made up of particle sizes suitable for the building element for which the mix will be required. Details on the analysis of sand are given in Section 4.3 and shown in Figures 4.40 and 4.41.

5.5.4 Test moulds

Prepare moulds for producing compact test samples that can be used in field tests to establish the satisfactory stabilization, compressive strength, and permeability of trial mixes.

Block moulds. Large, full-size moulds are only needed for testing bricks, blocks, and possibly foundation and sub-floor mixes. Block moulds need to be sufficiently strong to withstand compaction of soil and repeated use. If available, a manually operated or motorized block press is the most convenient means of preparing brick and block samples to test both stabilization and strength. Block sizes should be based on those used locally or by a standard block-making machine. The CINVA ram press, for example, produces 290 mm × 140 mm × 90 mm blocks (slightly smaller than 12″ × 6″ × 4″), which is a size frequently used. Box moulds for test mixes can be made with timber, provided they are well made and the inside surfaces are smooth. Plywood, particularly marine ply, is ideal for this purpose if it is available. (See photos 17a and 17b.)

Test cube moulds. Full-size block moulds are not needed when testing a range of mix proportions for stabilization. The amount of material and time taken for testing can be reduced by using small 50 mm × 50 mm × 50 mm test cubes. Box moulds for these are best produced as a three-gang mould – a set of three adjacent cubes (see Figure 5.7 and photo 17b).

Disc moulds. Another method of making small samples for immersion and permeability field testing is to cut 75 mm (3″) or 100 mm (4″) diameter plastic pipe into 25 mm (1″) or 50 mm (2″) deep rings for moulding discs of the trial mix.

A standard thickness of sample should be maintained for the comparative testing of cured mixes. If the thickness of the finish in the main work is known (e.g. for renders or screeds), this can be matched by that of the moulds for samples to be tested. 25 mm (1″) is a convenient thickness for this test.

Test-mix samples should be very carefully removed after initial curing of, say, seven days, and, to assist with this, the internal mould faces need to be smooth and lightly oiled. Continue to cure the test-mix sample outside the mould. Allow a full 28 days for the curing of all samples before testing.

5.6 Selecting the form of lime

5.6.1 Effects of lime on soil

The three main forms of non-hydraulic lime – quicklime, lime putty, and dry hydrate – can all be used to stabilize soil for building elements. However, when all forms of lime are available, there will be a best option since soil grading, plasticity, and liquid limit will be affected by the form of lime used.

- *Lime putty* of the correct consistency will have a small effect on moisture content, and the mix may need little or no additional water.
- *Lime dry hydrate* will dry the tempered soil and additional water will be required.
- *Quicklime* will slake, heat, and dry out tempered soil, producing a mix that will require the largest amount of additional water. (Quicklime is used to dry out waterlogged soils for this reason.)

Best quality powdered quicklime, used with sufficient water, is generally likely to give the best results for most stabilized-soil building elements, although well-matured lime putty is preferable for stabilized-soil plaster finishes when best quality quicklime powder cannot be manufactured, or is difficult to produce.

Hot-mixing with best quality quicklime powder is the most efficient method of stabilizing clay soil, although it must be handled with care. It is not advisable to use unslaked quicklime lumps or granules in renders and plasters, as the quicklime may continue to slake in the mix after it has been applied, which causes 'pitting and popping' of the surface, spoiling the finish.

Dry hydrate is a secondary choice, and could be used instead of lime putty if it is not possible to slake the lime to putty; or in place of quicklime if it is not possible to crush and sieve best quality quicklime. The hydrate should first be put through a fine sieve (Table 4.1) to eliminate unslaked particles which would produce defects at a later date. If the hydrate cannot be used fresh from the kiln, it should be kept in airtight containers. All dry-stored lime deteriorates with age, as it absorbs carbon dioxide and loses moisture, whereas, conversely, non-hydraulic lime putty stored under water improves with age.

5.6.2 Stabilization with different forms of lime

Quicklime. When stabilizing clay soils, it is best to crush and finely sieve quicklime for use in all building elements, including foundations and block mixes. It is preferable to do this mechanically rather than by hand, to reduce health risks from the quicklime dust. Very finely crushed, best quality quicklime can be used instead of lime putty in render and plaster backing/levelling coats. Quicklime is more reactive than lime putty, but if used for render, it must be very finely powdered, of the highest quality and mixed with the soil to an even colour and consistency. Speed of set will vary depending on the quality of materials. Some lime-stabilized renders prepared this way will need to be applied immediately after mixing, and others may take time to mature. The best approach can be determined by field testing and preparing test panels (see Sections 6.8. and 6.9). Major advantages of using quicklime in the mix, provided it is of sufficient quality, are that it acts rapidly and can prevent shrinkage cracks forming during set.

Dry hydrate. This is useful when it is too difficult or not possible to crush the freshly burnt quicklime, possibly due to lack of tools or appropriate personal protective equipment (PPE), or because the quicklime is too hard to crush (which would indicate it may be of poor quality). Dry hydrate can be produced quickly from quicklime and dry sieved immediately. It is also convenient for ease of packing and transport in bags.

Lime putty. After being stored under water, this is usually of better quality than dry hydrate and may be used with clay content or sandy soil, or with sand and pozzolan-based renders and plasters. The putty provides a stickier mix and good workability. Putty production requires more time and equipment than crushing quicklime or slaking to dry hydrate. However, lime putty is an excellent method of producing best quality lime and offers a means of long-term storage of fresh lime if the quicklime cannot be crushed and used immediately. As explained in Chapter 4, lime putty stored under water improves with age.

For initial trials, as a rule of thumb, the amount by volume of dry hydrate or lime putty that may be required in a mix is about double that of powdered quicklime, mainly because of the swelling that occurs during hydration.

5.6.3 Stabilization of clay soil with quicklime

As outlined above, because best quality, finely sieved quicklime powder is the most efficient and effective way to stabilize clay soils and can reduce or eliminate cracking, it is highly valued but needs to be used with caution.

Mix fresh, fully reactive, dry, crushed, and sieved quicklime powder directly into well-tempered clay soil (or sand and pozzolan). This is known as hot-mixing. The mixing can be done either starting with dry materials first, adding water at the end, or with pre-wetted (damp) soil or sand. It is best to use soil that is well wetted at the start, but not too wet on completion of slaking. This is because quicklime will cause a wet mix to dry out and stiffen up quickly, possibly in as little as 15 to 30 minutes, so it is important to add the right amount of water for the finished mix to be damp, but not saturated. This may take a little practice.

Figure 5.9 Hot-mixing – add the minimum water needed to ensure all lime is slaked and the mix remains damp.

A hot mix with finely powdered quicklime and soil should be used almost immediately after it has been thoroughly mixed to a soft workable consistency. If the quicklime powder is not sufficiently fine, the lime will take longer to slake and the mix should then be allowed to mature (mellow) for a day or more (possibly up to two weeks) until it reaches an even colour after remixing. Two days, however, is the maximum mellowing period normally recommended if all the quicklime has fully slaked in the mix during this time.

(See photos 4a and 4b.) The mix must not be too wet or too dry, or there will be a loss of strength of the finished work. Be prepared to add more water to bring the mix to a workable consistency. Any mix allowed to mature for any length of time should be 'knocked up' again (remixed) immediately before use.

There are added health and safety risks with hot mixes. Protect your eyes and skin, and wear a dust mask when using quicklime, especially when powdered. Use shovels to mix – do not expose bare skin to it (see Figure 5.10).

Figure 5.10 Use shovels for mixing lime. Do not work with bare hands or feet.

Good compaction will improve the strength of the finished work, but before final placing and compacting, ensure all quicklime in the mix has fully slaked and that there are no unslaked lime particles left (all the small, sieved lumps of quicklime should have broken down).

Test sample mixes for optimum preparation times, using additional samples if feebly hydraulic and/or slow slaking lime is used.

Figure 5.11 The starting point for mix proportions for lime stabilization is set out in the tables in Section 5.4.1, but the final proportions of materials for mixes (and various methods of mixing) will be determined by successful Stage 3 test results.
Use Stage 2 tests to establish the optimum proportions of lime to soil for stabilization. Keep to these proportions when making further modifications to mixes (e.g. by adding materials such as sand, fibres, and gravel) in order to make them suitable for different building elements.

5.6.4 Stabilization of clay soil with lime putty

If it is not practical to produce best quality, finely powdered quicklime, lime putty is an excellent alternative, provided it is of the quality recommended and verified by the Stage 1 field tests. Lime putty of this quality may be used to stabilize all building elements. It may be more manageable for mixing earth plasters and renders, particularly those made with sandy soil or sand and pozzolans: that is, mixes with none or little of the sticky, binding qualities of clay (see Figure 5.12).

The volume of putty, however, may need to be up to twice that of quicklime to achieve similar stabilization (see Table A1.3). Ensure all mixes with lime putty and all other ingredients are uniform in colour before they are used.

Figure 5.12 Lime putty can be used to stabilize all building elements and may be more manageable and workable than other forms of lime with mixes containing larger amounts of sand, or sandy soil, and pozzolans.

Mixes using lime putty will benefit from thorough mixing sustained for a minimum of 20 minutes. Mixes with clay soils may then be used immediately or allowed to sit in damp conditions under shade, for several hours to a full day or more, before being mixed again immediately prior to use. Keep mixes damp and covered, and do not allow them to dry out at all before final use (see Figure 5.13).

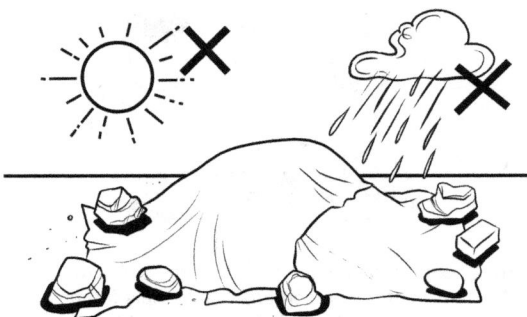

Figure 5.13 Mixes with lime putty and soil are best left covered and kept constantly damp. They can be stored this way for 24 hours or more, and then mixed thoroughly again prior to use.

5.6.5 Stabilization of clay soil with dry hydrate

Another alternative to quicklime is lime dry hydrate powder, although, as with lime putty, a greater proportion by volume – up to double that of quicklime – will be required to produce similar results and stabilization (see Figure 5.14 and Table A1.3).

Figure 5.14 Stabilizing with dry hydrate.

Fresh dry hydrate may be readily available when quicklime powder or lime putty is not. It can be combined with soil in a process similar to hot-mixing, outlined above. Ensure the final mix is of a uniform colour and is kept damp until used.

5.6.6 Stabilization with lime and pozzolan

A soil's low clay content may sometimes be supplemented by the addition of a pozzolan (see Section 4.4). Burnt clays and various types of ash react with lime and may (subject to testing) enable it to remain set under water.

Crush a sample of fired clay from broken bricks, tiles, or pottery into dust, or obtain the dust from a brick works (make sure that it is just brick dust and not contaminated with other materials), or use another pozzolanic material. Usually, the finer the pozzolan the more reactive it is, so use only the dust that will pass through a very fine mesh sieve – possibly 0.18 mm (ASTM No. 80) or even 0.063 mm (No. 230). Wet sieving will be necessary for the finest material. For practical field tests, 3.35 mm (No. 6) or 0.85 mm (No. 20) mesh sieves could be used on the understanding that the finer dust will act as the pozzolan and the larger particles mostly as aggregate.

Most traditional, manufactured pozzolans (largely burnt clays) were fired at between 700°C (1,300°F) and 900°C (1,600°F) before crushing to powder, which should be enough to ensure they are sufficiently reactive, although furnace ash and natural pozzolans such as volcanic ash were produced at higher temperatures.

114 BUILDING WITH LIME-STABILIZED SOIL

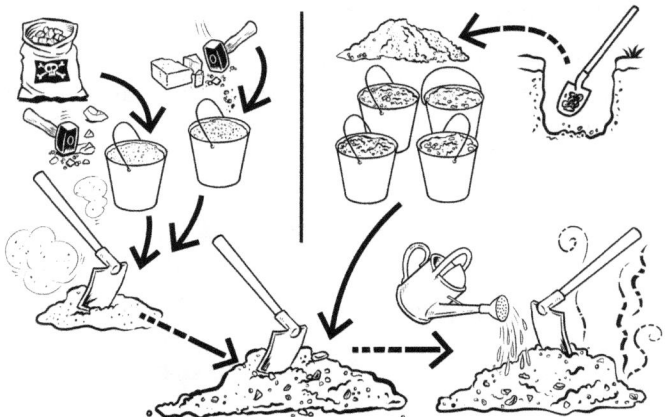

Figure 5.15 After testing, finely sieved pozzolan can be added to low-clay-content soil and lime trial mixes, or to sandy soil mixed with lime and other aggregates, to improve hydraulic set.

The amount of pozzolan that needs to be added to a mix to ensure stabilization depends on its reactivity (see Section 4.4). It can be added to lime in all its forms (quicklime, lime putty or dry hydrate), and the optimum ratio for a lime–pozzolan only mix – which will possibly be in the order of two or three parts crushed brick or burnt-brick dust to one or two parts lime putty – can be confirmed by testing. Pozzolan can also be added to a lime–sand mix (say two parts lime to one part pozzolan) to provide a hydraulic set. These mix ratios vary widely depending on the pozzolan's reactivity and the degree of hydraulic set or hardness required for the finished work.

Carry out the immersion test on specimens containing different proportions of pozzolan after 28 days curing, to establish the optimum mix for the hydraulic set and compressive strength required.

5.6.7 Un-stabilized mixes

High-quality plasters and renders can be produced with lime putty and sand only, or sometimes with lime and sandy soil. Without the addition of pozzolan, these finishes are un-stabilized and are therefore more appropriate for sheltered areas well above flood level where conditions are dry, or for internal plasterwork.

It is possible to create high-quality, thin, lime finishing coats for use on stabilized and un-stabilized soil backgrounds using lime putty and fine sharp sand, marble dust, or other clean, fine, and sharp aggregates. These can be further improved with polishing. However, mixes of lime and sand only require a far greater proportion of lime than lime-stabilized soil mixes: generally in the order of 1:3 and up to 1:1 lime to sand. These lime-rich mixes also need to include fibres to reduce the possibility of cracking, particularly for the thicker coats.

Un-stabilized mixes are not suitable for use in places with prolonged exposure to floods, or wet or damp conditions. A well-prepared non-hydraulic lime–sand render that has been correctly applied and allowed to fully carbonate and dry will withstand wet conditions for short periods, but not continuous saturation or frequent soaking.

5.6.8 Compaction

An essential part of preparing sample mixes for testing is to ensure maximum compaction of the material in the mould. This should be achieved using the same compaction method as for the main work, although in most cases it will be on a smaller scale. Subject to location and conditions, the mix could be compacted by hand using a rammer, by a hand-powered compaction machine such as a CINVA or similar ram for bricks and blocks, or by a more automated mechanical method. As a guide to the extent of compaction required, John Norton (1997) recommends that the volume of loose uncompacted soil should be 1.65 times greater than the volume it occupies when fully compacted in a block. This is a general guide that will vary with soil types.

Mortars, plasters, renders, and screeds are compacted by hand when they are placed, and the compaction is increased by heavy scouring and polishing.

Small hand tampers or pestles (see Figure 4.3) can be used for cube and disc samples. On removal from the mould, check to ensure the samples have been adequately compacted: the texture of the sample should be consistent on all sides and particles should be closely packed and evenly distributed without voids. National standards recommend that 50 mm (2″) test cubes are made by compacting three equal layers of the mixed material with the equivalent of up to 50 blows with a hand rammer. Square rammer ends should be used with square moulds to ensure that the corners are adequately compacted.

5.7 Curing

5.7.1 Tending for carbonation and chemical set

After preparing different trial-mix samples, based on clay content and suggested proportions of lime to soil given in Table 5.4, it is important that all samples are properly cured before testing.

In the last stage of the lime cycle (see Figure 3.1), lime absorbs carbon dioxide and loses water, reverting to calcium carbonate when it is fully cured. Reabsorption of water assists both the carbonation and the chemical set of hydraulic mixes: high temperatures and the presence of moisture assist and lead to the production of hydraulic compounds that remain stable under water. The more often mixes are dampened, the water allowed to slowly evaporate, and the mixes dampened again, the faster the curing is likely to be, and the more successful the set. The chemical (hydraulic) set can continue in permanently wet conditions.

It is therefore essential that carbonation and set are assisted by damping down, particularly in extreme temperatures and dry climates, if the mix is to achieve full strength and perform at its best. If mixes are well shaded, particularly in hot climates, they are less likely to dry out too quickly. If they are not tended, and dry out too rapidly, there is likely to be little or no increase in strength.

5.7.2 Curing procedure

Fully cure all trial-mix samples by carefully placing them on a flat surface under shade, and keep them damped down for 28 days. Ambient humidity and temperature will determine how often the samples should be dampened, but in hot and dry conditions, it is recommended that damping down should take place no fewer than three times a day, preferably more. Ideally, the samples should be dampened as often as possible, depending on how quickly the mix dries out. As explained above, this can increase the rate of carbonation and speed of hydraulic set during the curing process. Ensure that someone responsible is curing the samples, and that the frequency and extent of damping down are accurately recorded. Damp sacking or cloth draped over finished work and kept damp is another good method of curing and protection.

5.8 Field testing trial mixes

5.8.1 Preparation

It is important to carry out the following field tests on trial mixes after 28 days curing so that the optimum mix design for use in the main work can be confirmed. The immersion (soak) test is the most important of these as it establishes whether the soil has been sufficiently stabilized. (See photo 17c.)

5.8.2 The immersion test (soak test)

No mix can be considered to be fully lime-stabilized without successful immersion and wet compressive-strength test results. Good results confirm that the specific lime–soil mix is suitable for use and will remain stable under water. The immersion field test is important for all trial mixes designed to withstand wet conditions or be flood resilient. This will confirm which lime proportion has successfully achieved stabilization of the soil.

Figure 5.16 The immersion test assesses the durability and stability of treated soil for wet and flood conditions.

- Fully immerse all cured samples (blocks and mortar cubes, render, plaster, and floor-screed discs) in water.
- Keep them under water for the maximum anticipated length of continuous wet or flood conditions, or for as long as possible.
- Monitor regularly to check on stabilization and strength: no sample, or part of a sample, should have dissolved in the water, and the sample should remain sufficiently strong in compression. All results should be clearly recorded. (See photos 15a and 15b.)

If, after the required period of time, all samples have remained stable under water, consider making further trial mixes with less lime, to establish the minimum amount of lime required for full stabilization. Alternatively, to speed the process, a greater number of samples, with a wider range of lime proportions, can be made and tested at the same time.

It is not always the case that the higher the lime proportion in a mix, the greater the strength or stabilization. Research has shown that, in some cases, a lower lime proportion will stabilize a mix more effectively than a higher one (Little, 2000). The optimum lime proportion for stabilization of different soils and mixes varies, which is why preparing a range of field-test mixes is advisable.

5.8.3 Monitor results

The immersion test indicates how well a block, or any lime-stabilized building element, may last without dissolving in rain or flood conditions, under water or below ground. Keep an accurate record of the different mixes used, and ensure there are clear references to each mix's materials and ratios on the samples under test. (See Appendix 6 and Table 5.9.) No matter how short or urgent the construction programme, ensure these reference numbers are permanent and do not wash away while the samples are under water, as it is important to have a permanent and accurate record of successful mix ratios.

Test samples, including wall building blocks and foundation blocks, should remain under water for a minimum period of one month, preferably for longer. The best mixes could be tested for up to a year or more. Laboratory testing should verify that fully stabilized mixes will not dissolve at all. They should pass the wet-strength test: the compressive strength after 28 days curing should be at least 1.5–3 N/mm² (220–440 psi). Field tests for compressive strength are given below, but in addition to these, if possible, arrangements should be made for laboratory testing in accordance with ASTM standards (see Chapter 8), at two months, six months, one year, and two years. (See photos 17a and 17b.)

5.8.4 The step test

This is a simple field test to assess the strength of a block. Ensure the block to be tested has been properly cured for a minimum of 28 days before carrying out the test.

Using cured and dry blocks, before their immersion testing, place the block to be tested lengthways across a gap between two other blocks that are also placed lengthways. Ensure the top block overlaps the lower support blocks by no more than 50 mm (2") either side. The wider the gap between the supporting blocks, the more demanding the test: the gap should not be less than 200 mm (8") for blocks and 150 mm (6") for bricks. Find someone who is heavy and has good balance to stand on one foot in the centre of the block being tested so it takes their full weight (see Figure 5.17).

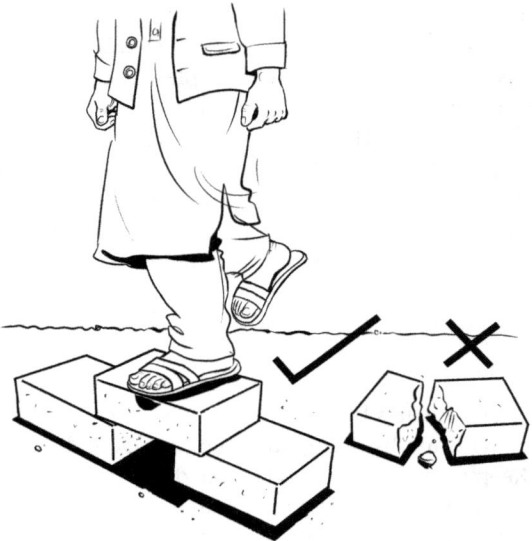

Figure 5.17 The step test.

If a common size block breaks when overlapping the supporting blocks as above, then it is not strong enough. It may have needed more damping down and time for curing, the materials may be defective, the mix ratio may be unsatisfactory, or it may not have been mixed well or compacted enough.

If trial mixes with lower proportions of lime pass these tests, it may be possible to make successful modifications to the mix and further reduce the amount of lime required, while retaining sufficient strength and stabilization. Indeed, reducing the lime proportion may, in some cases, increase strength as well as provide a cost saving.

If blocks start to dissolve in the immersion test, but were strong in the step test, they have not been adequately stabilized, and the lime to soil ratio of the mix will probably need to be revised. The reasons for failure should be established, the deficiency corrected, and further testing carried out. It may be that the proportion of lime in the mix needs to be changed (see Section 5.4) or that the clay soil is unsuitable for stabilization – although this should have been determined during the tests for soil suitability in Stage 1 of the field-testing programme. Check that all the parameters for all the materials successfully tested in Stage 1 are maintained before re-testing.

However, un-stabilized blocks that are strong enough in the step test should be appropriate for use in dry situations, or above the flood line. If they are finished with a stabilized render, and topped by a good roof and overhang, they will be protected against water penetration.

5.8.5 Compressive strength

John Norton (1997) suggests that the dry compressive strength of un-stabilized compressed mud brick is at least 1.5 N/mm^2 (220 psi) after one month of drying, depending on climate. Ugandan and Tanzanian building codes require compressive strengths of 2.5 N/mm^2 (363 psi) loadbearing walls and for one- and two-storey buildings, the Greater London Council requires 2.75 N/mm^2 (400 psi) (GLC, 1973). Arup Engineering advises a compressive strength of 5 N/mm^2 (725 psi) for loadbearing walls in seismic areas (Arup, 2017).

Compressive-strength values for stabilized soils are often given as wet strengths, since this is the critical strength condition. These values should be at least half the dry strength value. A suggested minimum wet compressive strength for internal partition walls is 0.7 N/mm^2 (100 psi) (Norton 1997).

Lime and lime-stabilized compounds develop their strength slowly, and to allow for this tests of each mix should be carried out at the standard 28 days, and also at two months, six months and one year. A particular mix may take two years or more to reach full compressive strength. However, a range of field tests on well-prepared lime-stabilized soil mixes in Pakistan in 2013 and 2014, using hand-held penetrometers, indicated compressive wet strengths of over 4 N/mm^2 (600 psi) after one month (28 days) curing, the maximum reading of the penetrometer. Similar strengths were recorded for mixes immersed in water for several months, some for a year or more, and such field-test readings continue to be recorded through on-going testing in various countries.

Wet compressive strength field testing of a variety of trial mixes for sub-road stabilization with the Technical Research Team at Tamera, Portugal (2018–2019), recorded wet compressive strengths of 7N/mm^2 and above, for a variety of cured trial mix samples after two weeks under water. An increase in compressive strength after one year's immersion in water was recorded for several of the same samples. Example mixes were: 1 part quicklime: 7.5 soil: 3 sand: 4 gravel and 1 part quicklime: 7 soil: 3 gravel.

Long-term strength gain of lime-stabilized soil has been known for some time and was investigated by the Texas Highway Department in 1966 in connection with road construction. Core samples were taken from ten-and-a-half-year-old lime-stabilized soil roads between Houston and Beaumont and were found to be 'approximately three times stronger than would have been predicted from ordinary, preliminary laboratory tests'. A further example of long-term strength gain is described in Section 7.2.6, where a compressive strength reached 12.8 N/mm^2 when tested after 25 years. (See also section 7.4.2, 'Final strength'.)

120 BUILDING WITH LIME-STABILIZED SOIL

Appendix 7 gives the compressive-strength requirements for the walls of domestic buildings stipulated in the Greater London Council's *London Building Constructional By-laws 1972* (GLC, 1973). It can be seen that field-test results demonstrate that the compressive strength of lime-stabilized soil blocks exceeds these requirements. (See photos 15a and 15b and Appendix 7.)

5.8.6 Water permeability

If sample discs of render and floor-screed mixes are uniform in size and depth (about 25 mm (1") thick), they can be subjected to comparative water permeability field tests to assess how long the samples will resist moisture penetration and saturation. (For vapour permeability, see Appendix 10.)

- Place each of the cured discs inside a standard funnel, seal the sides with a water-resistant sealant, and allow the sealant to set (depending on the type of sealant, this may take up to 24 hours). It should be possible to select from many of the low-cost sealants that are available 'off the shelf' from building merchants, or use beeswax or candle wax.
- Place the funnel in the neck of a dry jar (see Figure 5.18).
- Gently pour water onto the sample until it is fully submerged and the water above it is at least 75 mm (3") deep. Make sure the head of water above each sample is consistent and the same height as the water in the main work (i.e. that it is the depth of the water channel, container, or

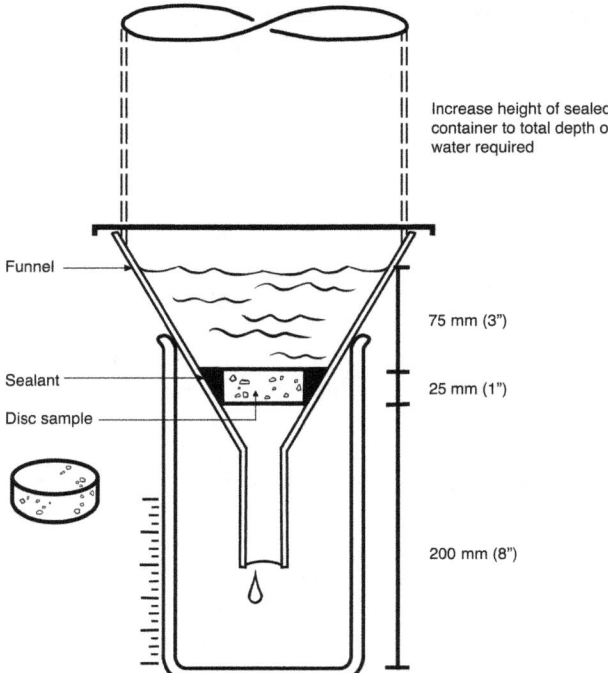

Figure 5.18 Water permeabilty field testing.

storage tank). If it is for checking water resistance to render, a 50–75 mm (2″–3″) depth of water should be sufficient.
- Monitor and time the extent and speed of any water penetration, and note the rate and duration of seepage over hours, days, or weeks to assess whether the mix is suitable for the intended location, and compare alternative mixes. The extent to which lime-stabilized soil mixes are able to retain water is remarkable.

5.9 Recording mix ratios

Ensure that every test sample is clearly marked or labelled with details of its mix and the dates it was made and immersed. Mix ratios can be scratched carefully into the surface of a sample before it has dried, and labelled in ink

Table 5.8 Suggested abbreviations for recording mix components.

Material	Abbreviation
SOIL	SO
LIMES	
Quicklime	QL
Lime putty	LP
Dry hydrate	DH
SANDS AND AGGREGATES	
River sand (soft)	RS
Hill sand (sharp)	HS
Gravel	GR
Broken stone	BS
POZZOLANS	
Burnt-brick dust	BBD
Rice-husk ash	RHA
Volcanic ash	VA
Pulverized fuel ash	PFA
Blast-furnace slag	BFS
ORGANIC MATERIAL	
Cow dung	CD
Tallow	T
FIBRES	
Hair	HA
Hemp	HE
Jute	J
Sisal	SI
Straw	ST
OILS	
Linseed oil	LO
Castor oil	CO
Palm oil (note significant environmental impact)	PO
Mustard seed oil	MO
Locally available oils	As appropriate

Table 5.9 Using abbreviations to record the composition of mixes.

Parts written as	Composition of mixture
1QL : 10SO	1 part quicklime, 10 parts soil
2LP : 15SO: 2BBD : 3J	2 parts lime putty, 15 parts soil, 2 parts burnt-brick dust, 3 parts jute
2DH : 16SO: 2RHA	2 parts dry hydrate, 16 parts soil, 2 parts rice-husk ash

with a permanent marker. When immersion testing several trial mixes at the same time in one container, ensure that each sample is always identifiable. For example, a mix written in permanent marker onto the surface of a block, cube, or disc may eventually become unreadable if the surface discolours or erodes beyond a few millimetres. Consider placing the clearly marked sample discs or cubes in transparent bags that are themselves filled with water and labelled with permanent marker, and immersing these in the container. This helps avoid confusion and also helps to fit more samples into one container. It may be necessary to re-mark the bags with permanent marker each month, to avoid the loss of the label over time.

Mixes are generally written as a ratio showing the number of parts, starting with lime, followed by soil. The order could continue with sand, then pozzolan and finally organic materials such as fibre and cow dung; the last of these will depend on whether additives are used to further modify the mix to make it suitable for a particular building element. Abbreviations, such as those listed in Table 5.8, can be used to make recording easier. (The list in the table is not exhaustive and other items such as marble dust or grit can be added.) Some examples of how these may be used are given in Table 5.9.

5.9.1 Test-mix record format

Before undertaking tests, prepare all recording material including permanent markers and test record sheets (see Appendix 6 for an example). To benefit from these tests, meticulous recording at all stages is essential. Details should include date, location, depth of soil extracted, type of test undertaken, duration, result, and name of tester. Particular attention should be given to the mix materials, proportions, and quality of the lime used. Record the exact source of the soil and other materials. Record curing conditions, monitoring time, and outcome of immersion tests. These records will assist in the preparation and reproduction of successful mixes. It is helpful to support records with photographic documentation.

CHAPTER 6
Field testing: Stage 3 – Building elements

6.1 Production methods

6.1.1 Testing final building elements and components

Stage 3 of the field-test programme is to test finished components and elements made using materials and mixes based on successful Stage 1 and 2 test results. The bulk material to be supplied for the main work should be used for these tests before commencing full production and construction. Check that the bulk soil supply for the main work is identical to the soil samples used for testing mixes in Stage 2, and the production process gives the same satisfactory results. The proportion of lime to soil needed for stabilization will remain as determined in Stage 2, but the variations required for each building element should also be assessed. Adjustments to mixes (fine-tuning) in the form of variations in the proportions of fibres and aggregates is likely to be necessary for different components. For example, the tensile strength of plaster can be improved with the addition of fibres; areas subject to heavy wear or requiring greater water resistance may be improved with additional pozzolan; and foundation and floor-slab mixes may benefit from larger aggregate.

Building elements produced for the main work should be tested during manufacture for consistency of quality, and before construction. Tests should be carried out on all elements at regular intervals: for example, for block-making, test at least three, and preferably five, of the first 500 blocks made. Test mixes at intervals, for stability and both dry and wet compressive strength, for at least two years. Building elements made by hand may not achieve the consistency and compaction possible with mechanized processes.

6.2 Foundations

6.2.1 Trench footings with lime-stabilized soil or limecrete

Strong foundations are important for all buildings and essential for construction in wet conditions. In long-standing flood water, the foundations are likely to be saturated for longer than other building elements. All materials and mixes used for foundations, whether lime-stabilized soil, limecrete (see Glossary), block, brick or stone, and mortar therefore need to be thoroughly tested to ensure that they remain stable under water.

6.2.2 Excavation

The shape and size of foundations will vary with local conditions. When in doubt about foundation design, the advice of an engineer with practical experience of earth building should be sought regarding, in particular, compressive strength, depth, and width. As a general rule, after removing the topsoil, the foundations should be taken down to a solid base. The depth of this solid base or ground beneath the surface varies. Ground conditions in towns and villages should be carefully examined for accumulated landfill, the presence of which should be avoided or taken into account in the foundation design. Recently disturbed land, including that used for agriculture, is similarly suspect, and foundation design and location need to be planned accordingly.

Dig the foundation trench to the same size as the foundation required to support the walls and all the superstructure. Compact the loose ground at the bottom of the trench by ramming and tamping it down well. Damp down the bottom and sides of the excavation before placing foundation material. Trench footings are usually at least twice the width of the walls they support (subject to the engineer's advice on individual site conditions).

Stage 2 field tests will have established the lime to soil ratios necessary for stabilization. These proportions should be retained, but for foundations, large aggregate can be added subject to further testing of the modified mix. If solid, strong and inert materials – such as rock, stone, large gravel, or broken brick – are available, these can be incorporated in trench foundations as aggregate provided that they are well broken-up, compacted, and bound together with fully stabilized soil or a hydraulic lime mortar. Larger aggregate should still be well graded and have a minimum size of about 25 mm (1"). Mix variations like these, incorporating more or larger aggregate, should be immersion tested before use.

Figure 6.1 Digging the foundation trench and compacting the ground.

For examples of foundation options, see Sections 6.2.4 and 6.2.6 and Figures 6.5 and 6.6.

6.2.3 Appropriate water content

The minimum amount of water required to obtain a plastic mix should be used for stabilized-soil trench foundations and footings. If using quicklime, be aware that quicklime dries out the mix, so more water will be needed if using this than if using lime putty or dry hydrate, and additional water will need to be added during mixing to keep the mix plastic.

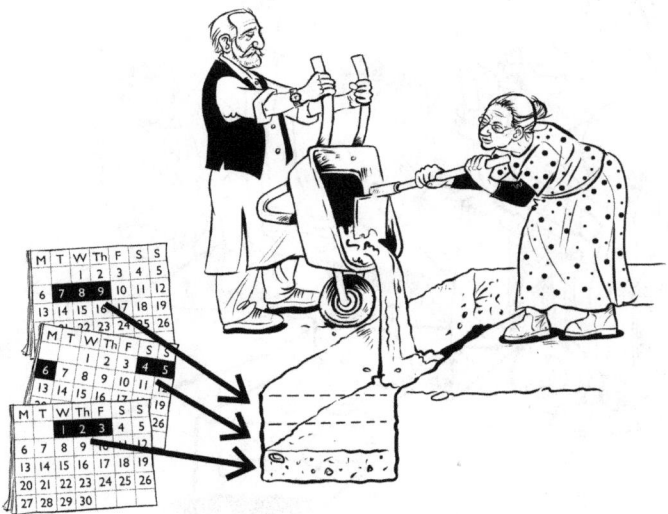

Figure 6.2 Laying the foundations with a mix that is workable and plastic but as firm as possible to enable good compaction.

Once the mix has an even colour and plastic consistency, use the minimum additional water but add sufficient to keep it workable. The mix must be damp but firm enough to be compacted to hardness immediately after placing. If it is too wet, it will be weakened and good compaction will not be possible. This point also applies to all other compacted-earth building elements, such as compacted blocks and rammed-earth walls. Quicklime, however, dries out soil rapidly, and considerably more water will be required to maintain a mouldable consistency when this form of lime is used.

6.2.4 Compaction

It is important that a stabilized-soil foundation mix is well compacted, so the foundation trench should be shaped to accommodate this (see Figure 6.3).

Trench footings are regular in shape, and an advantage of compacted stabilized-soil trench footings is that the excavation needs no further work or backfilling. In addition, the sides of the trench act as shuttering. It is essential that the mix used for lime-stabilized soil foundations passes the immersion and strength tests before use in the main work. Build the foundation up in layers about 100 mm (4") deep and compact and keep damp, one full layer at a time.

Figure 6.3 Good compaction is essential for foundations.

6.2.5 Stabilizing different soils for foundations

Different soils should be stabilized in different ways, as summarized below, and all cases are subject to satisfactory Stage 2 test results.

Clay-rich soils. Use quicklime powder or granules to stabilize soil for use in foundations and floor slabs. Dry hydrate powder or lime putty may be used as an alternative, but stabilization with these is unlikely to be as fast or as effective. A larger volume will be required than if quicklime is used (see Table A1.3).

Clayey soils with a little sand. These may be stabilized in the same way as clay-rich soils. For foundations, they might be improved by the addition of sand, gravel, and stone aggregate, subject to testing.

Sandy soils. Use pozzolan and lime, or hydraulic lime mixes, to improve hydraulic set and water resistance. Strong hydraulic sets are required for foundations in wet or flood areas.

Gravelly clayey soil. As for clay-rich soils; use quicklime powder.

Gravelly sandy soils. As for sandy soils; use hydraulic lime or lime with pozzolan.

Silty soils. High-silt soils are not suitable for use as a building material. Very silty soils cannot be used without major modification and the addition of sand and/or clay or pozzolan. If soils that include high proportions of silt are modified, ensure mixes are thoroughly tested and only use them if successful results are reproducible. As a general guide, a soil's silt content should not exceed 20% for modified and stabilized earth mixes, and 6% for lime–sand (and lime–sand–pozzolan) mixes.

Large aggregate. Subject to local availability and cost, compacted broken-stone aggregate or dimension stones, or bricks or blocks in a hydraulic lime mortar, are possible alternative foundation materials (see Figure 6.6).

6.2.6 Suggested foundation options

A number of materials are suitable for building foundations but various forms and construction methods appropriate for the chosen material need to be considered. Materials most likely to be locally available are lime, soil, fired brick, and stone (see Figure 6.4).

Compacted lime-stabilized soil foundations using clay soils. Using a clay soil which also includes sand and gravel as the basis of a lime-stabilized soil foundation is likely to give the best results. Either mix the ingredients

Figure 6.4 Locally available materials suitable for incorporating in foundations in many parts of the world for a hydraulic set are: lime, clay soil and sand and aggregates; or lime, pozzolan, sand and aggregates, brick and stone.

dry first, or mix granular quicklime into damp, tempered, and pre-mixed soil, adding just enough water to achieve workability. All the quicklime granules or powder used should be fully reactive and fresh from the kiln. Consider adding more, and larger, gravel subject to availability and further testing. Mix very well until an even colour is achieved. Before the main work, test a block made with the mix for at least one month to ensure it is stable under water. Also test the effect of tempering the final mix for one or two days before placing and compacting. Table 6.1 gives the composition of lime-stabilized soil limecrete mixes that remained stable after six months immersion as examples. When tested, the strength of all samples was found to be in excess of 4.7 N/mm^2 (700 psi), which was the maximum the hand-held penetrometer could record.

1. Following successful foundation mix test results, place and compact the mix in the trench in a series of layers, and ensure each layer is compacted well. The layers should each be 100 mm to 150 mm (4″ to 6″) deep when compacted. Quicklime will dry out the mix and more water will be needed to bring the mix to a damp and mouldable but firm enough consistency to compact it. Pre-test the effect of tempering and stabilizing the final mix for different lengths of time before placing, and compacting, in the foundation.
2. Finish the top layer flat, leaving a level base on which the wall is to be built. Shuttering could be used to take the side of the foundation trench up higher, to form a plinth at the base of the wall where additional protection is often required.
3. Cure the lime-stabilized soil concrete foundations for a minimum of two, and preferably four, weeks. Protect them from hot sun and rain. Keep them humid by covering with wetted sacks (or tarpaulin) and damp down regularly.

Stone or brick foundations with hydraulic lime mortar. If strong stone is available it will make a good foundation, particularly if the

Table 6.1 Three examples (from hundreds) of successful lime-stabilized soil trial-mix samples remaining stable under water after six months immersion, tested by the International Organization for Migration (IOM) implementing partners (IPs) in Pakistan in 2014–2015.

Village	IP and district	Mix	Date of immersion of cured sample
Habibullah Chachar	AHD – Ghotki	1 Lime putty : 2 Soil : 2 Hill sand : 4 Gravel	20.09.2014
Kamoo Shar	CRDO – Ghotki	1 Lime putty : 2 Soil : 1 Hill sand : 3 Gravel	23.08.2014
Daro Ahmed Khan	SOD – Ghotki	1 Lime putty : 2 Soil : 2 Hill sand : 4 Gravel	23.08.2014

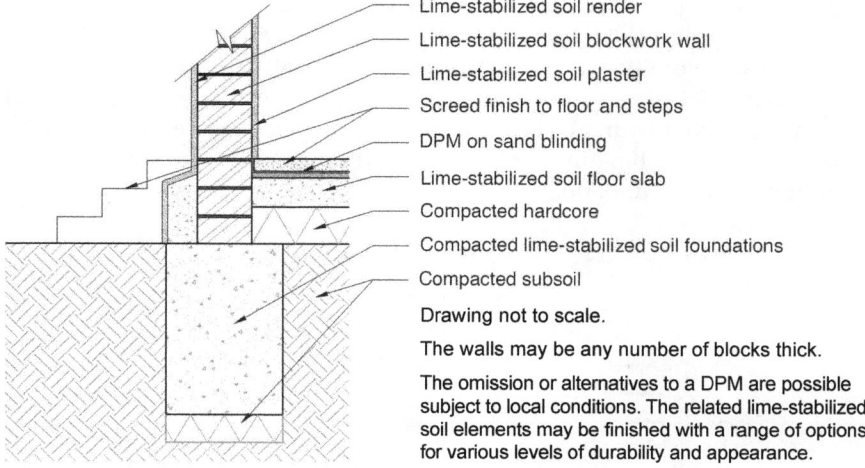

Figure 6.5 Foundation option 1: compacted lime-stabilized soil trench footing.

stones (or bricks) are mortared together. These mortars can be made with a hydraulic lime and sand, a non-hydraulic lime, pozzolan and sand mix, or a lime-stabilized soil mortar.

Brick, block, or stone foundations (see Figure 6.6). Un-stabilized earth bricks or blocks should not be used for foundations where there is a flood risk or saturated subsoil. Bricks, blocks, and mortars of lime-stabilized soil should only be used after full testing to ensure that the mix will remain stable under water and pass the wet-strength test. This also applies to units made of any other material, including fired brick.

Engineers' immersion tests in Northern Sindh in 2014–2015 compared lime-stabilized blocks with Class B fired-clay bricks. (These bricks are softer than Class A bricks, and slightly cheaper, so are commonly used for building in villages, including those in flood-affected areas.) Class B bricks lost 30–40% of their mass through dissolving in water after only five days immersion. The lime-stabilized blocks, however, remained fully intact and stable under water for many months.

Unlike trench footings, foundations built using small units such as bricks require a wide excavation in order to spread the load evenly, for adequate stability and to provide sufficient room in which to work. After building the foundation, the excavation has to be backfilled. Attention then needs to be given to avoiding subsidence of backfill resulting in uneven levels or depressions which would encourage water to remain close to the base of the building. It is important to ensure that final compaction of the soil and finished falls encourage water to quickly drain away from the base of the building. The additional work of backfilling is not required for trench footings.

Consider using raised ground levels, good falls, bund walls, and drainpipes or channels to protect the building and encourage rapid water run-off.

130 BUILDING WITH LIME-STABILIZED SOIL

John Norton (1997) recommends that unless specific requirements suggest a different approach, a foundation composed of individual units should spread out to an angle of 60° at its base, tapering to the wall's width at the top over a minimum of three or four courses. The courses must be bonded together well, and laid in a hydraulic mortar. The advice of an engineer with practical experience of earth building should be sought where there is any doubt about wall thickness or foundation design.

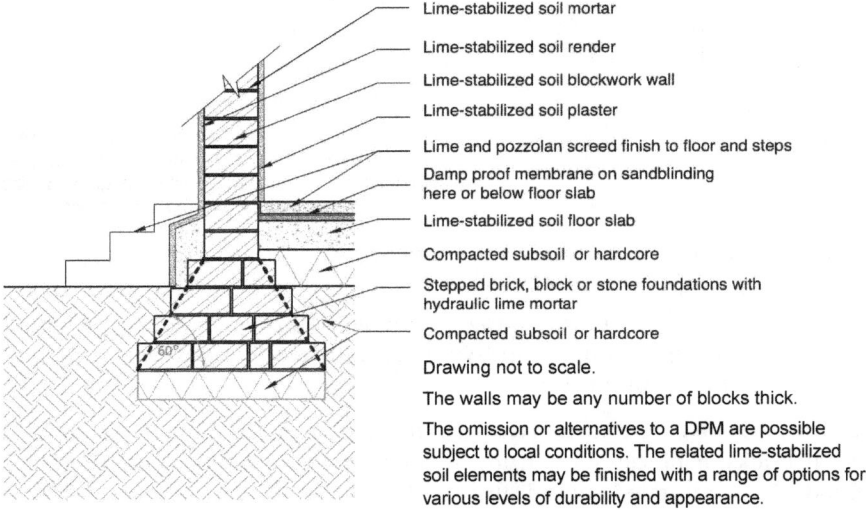

Figure 6.6 Foundation option 2: stepped block, brick, or stone foundations and related elements incorporating lime.

Hydraulic lime concrete for sandy soils. When designing a mix for lime concrete foundations using sand or sandy soil, mix the dry ingredients first and allow – subject to satisfactory Stage 2 test results – for three to four parts pozzolan (such as brick dust) to two to three parts lime putty (or one to two parts quicklime), four parts sharp coarse sand or soil, and six to eight parts gravel. The mix should be adjusted as required subject to the results of field tests. If crushed stone is available, this could be added as an alternative to the gravel. Bring all to a workable consistency, place in the trench and take care to compact and dampen each 100 mm (4″) layer (described in Section 6.2.4). Laboratory test this mix prior to use if possible. If available, a hydraulic lime could be used instead of a pozzolan and non-hydraulic lime mix.

Where regular flooding occurs and the maximum height of water is known, a protective 'toe' or plinth of lime concrete or lime-stabilized soil as a design feature could provide additional protection at the base of the wall. This may be made from a hydraulic lime or lime-stabilized soil concrete (shuttering may be required), or of rendered lime-stabilized blocks. The protective plinth can be added once the walls are built and cured, or built during construction of the walls (see Figure 6.7). Note that 'toes' or plinths may not be necessary if all building components above the foundations are fully lime-stabilized. However, large plinths could offer an additional protection from the risk of high-impact floodwater surges from hills or mountain run-off, and from events such as broken dams or river embankments during flood conditions.

Figure 6.7 Constructing a protective plinth using lime-stabilized soil or limecrete.

6.2.7 Curing

Foundations and plinths, like all lime-stabilized work, benefit from good curing conditions, which increase strength and assists hydraulic set. Keep all new works dampened and shaded from sun and rain for as long as possible, preferably not less than four weeks, before building off them.

Figure 6.8 Keep all foundation work well cured to ensure maximum strength and hydraulic set.

6.3 Bricks and blocks

The lime and soil proportions – finalized after successful Stage 2 testing – should be maintained in mixes for bricks and blocks. Lime-stabilized units of this nature may be moulded to any standard size or to a non-standard size to suit local custom. Each form of lime can be used for making them (quicklime, lime putty or dry hydrate). Hand-compacted blocks will require more fibre in the mix than machine-compressed blocks to minimize cracking and to strengthen them for handling before they are cured. When preparing mixes for hand moulding try adding fibre in the order of 10% or more by volume, subject to testing. If units are sufficiently compacted, they may be adequately cohesive without fibre (See photo 14).

When using quicklime for brick- and block-making, ensure that only well-burnt lime is used and leave the mix to complete the slaking process (check all the small lumps of quicklime have completely broken down) before compacting it in a mould. Provided the mix is kept damp, this slaking process can take anything from a couple of hours to a couple of days or more. Use a shovel or mechanical mixer to turn the mix over where preliminary dry-mixing with quicklime is preferred. The advantage of using quicklime powder is that it will disperse quickly in the mix, although dry-mixing tends to generate quicklime dust – which is caustic and can burn – so wet-mixing or using granulated quicklime may be preferable, particularly in windy conditions. Keep the mix shaded and damp at all times. Although the mix can be used immediately, try testing whether the mix is improved by tempering for a day or so, in which case be sure to re-mix (knock-up) the mix again before compaction in the mould.

An alternative to using quicklime is to use lime dry hydrate powder or lime putty, although a greater proportion by volume will be required to produce the same result, possibly up to double. The optimum amount of lime in a given form for stabilization can be determined by the Stage 2 field tests described in Chapter 5.

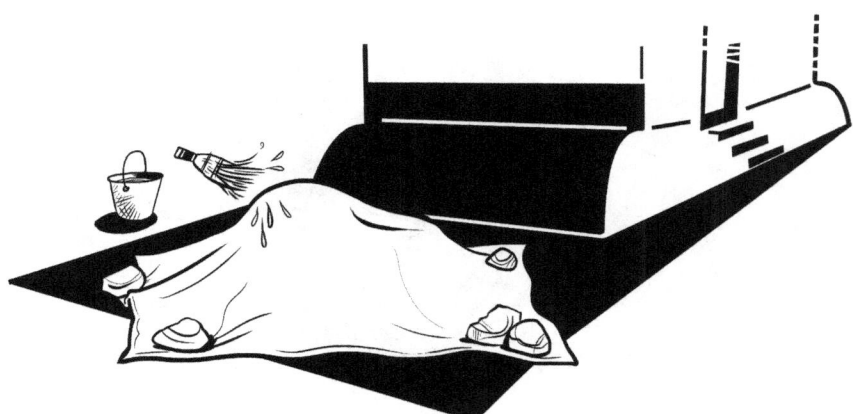

Figure 6.9 Cover mixes from drying sun or heavy rain if they are not to be used immediately.

6.3.1 Field tests for moisture content

To produce well-stabilized bricks and blocks with good compressive strength, use mixes with optimum moisture content (OMC) to ensure workability, and fill moulds with maximum compaction. Use the same sized containers to measure the water content of all mixes, and check consistency.

Ball-drop test for optimum moisture content. This is similar to the ball-drop test used to assess the approximate clay content of a soil, but in this case it is being used to check the moisture content of the whole mix.

Take a handful of moist block-making mixture and shape it into a compact ball. With arm outstretched, drop the ball onto a hard flat surface from a height of about 1.5 m (5'). If the dropped ball stays in one or two lumps, it is either too clay-rich or there may be too much water in the mix. If it breaks into four to ten pieces, the mix is probably too wet (or may be too dry with a low clay content). If it breaks into a few small pieces, but will stay together if squeezed hard in the hand, it is about right (see Figure 6.10).

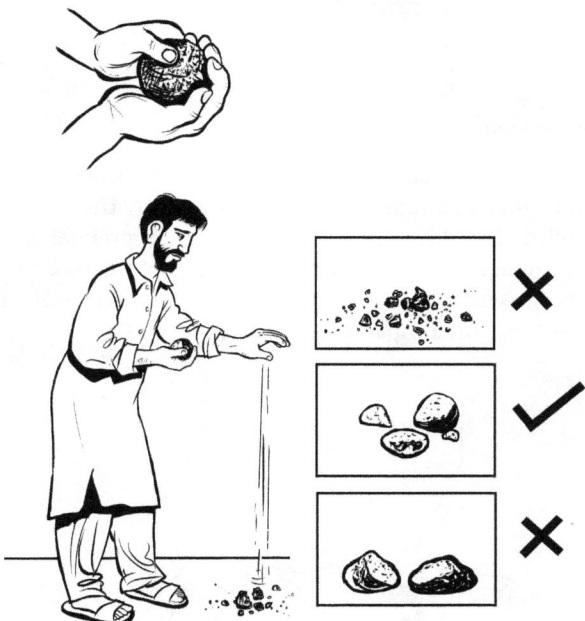

Figure 6.10 Ball-drop test for optimum moisture content.

Bar test. Place a large bucketful or shovelful of the uncompacted mix on the ground. Hold a 500 mm (1'8") long and 20 mm (¾") diameter steel reinforcing rod vertically, and loosely rest the end on the surface of the mix. Let the rod sink under its own weight. If the bar sinks in by exactly 20 mm (¾"), the water content is right.

6.3.2 Oiling the box mould

Oiling the inside surfaces of a box mould makes it easier to remove the compressed and moulded block by helping prevent it sticking or breaking (see Figure 6.11). An alternative method is to sprinkle very fine sand onto the insides of the mould before filling it.

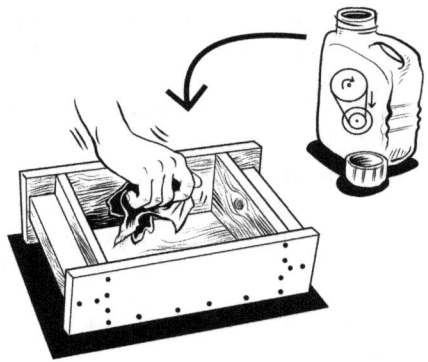

Figure 6.11 Oiling the block mould.

6.3.3 Filling the mould

Add sufficient mix when filling the box to ensure that the mould is completely filled with fully rammed material. Fill the mould with the mix, pressing it down firmly with gloved hands (see Figure 6.12) or, preferably, tamp the mix hard with a rammer, especially in the four corners of the box (see Figure 6.13). As a guide, the mould should eventually hold approximately 1.65 times its own volume of loose uncompacted mix (Norton, 1997).

Figure 6.12 Filling and compacting the block mould.

6.3.4 Compressing the mix

Good compaction improves the durability and strength of the block (see Figure 4.3 and Figure 6.13).

To achieve this for handmade blocks, use a rammer or a shaped piece of wood or stone on top of the mix and press or tamp it down hard. Ensure even pressure.

Figure 6.13 Compact the block mix well for best durability and strength, especially in the corners.

Stronger blocks can be produced using one of a range of hand or mechanically driven compressed-earth block-making machines such as a CINVA ram (see Figure 6.14 and Appendix 2). These machines produce greater compaction than is possible by hand due to the increased pressure from leverage and are widely available.

Figure 6.14 Hand-operated block press (CINVA ram).

6.3.5 Removing the block

Unless they are sufficiently well compacted to enable immediate demoulding, lime-stabilized blocks may need up to a week's initial curing before being removed. Take care that the block does not break or crumble when lifting it out of the mould (see Figure 6.15). To avoid disturbing the new block, it is best to carefully lift the box mould up and away from the block, and demould on top of a small supporting board which can then be used to carry the fresh block. Alternatively, compact the block directly onto a hard flat surface on which it may be subsequently cured without being moved.

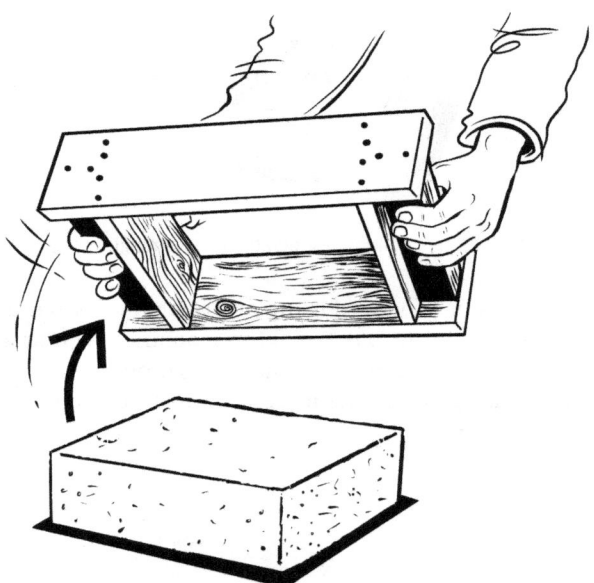

Figure 6.15 Demoulding the block – handle carefully.

6.3.6 Checking the block mix

If a block crumbles when carefully taken out of the mould (see Figure 6.16), this indicates that the block mix was too dry, or that it had not been adequately compacted or that the proportions of materials in the mix did not match those of a successful Stage 2 test sample. Check that all materials and mixes are the same as those that have been satisfactory in the Stage 1 and Stage 2 field tests and use for the following block-making method and Stage 3 testing.

FIELD TESTING: STAGE 3 – BUILDING ELEMENTS 137

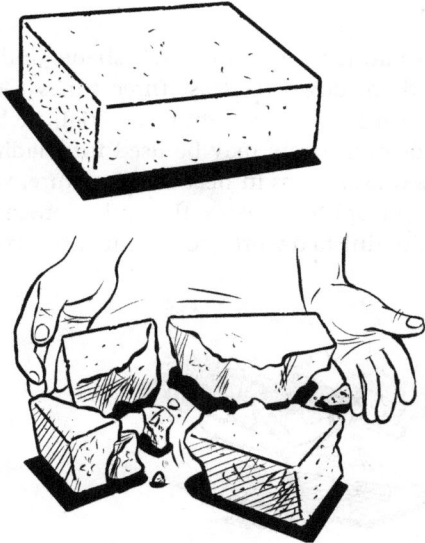

Figure 6.16 Testing the quality of a newly made block.

6.3.7 Care in handling the block

If demoulding cannot be done in situ, carrying the new block on a board will help to prevent damage (see Figure 6.17).

Figure 6.17 Careful handling and carrying of a newly made block.

6.3.8 Curing blocks

Place the blocks on flat, level ground in the shade, and keep moistened by regularly damping them down (at least three times a day) for four weeks or longer (see Figure 6.18). Any sheeting material, including sacks, cloths, plastic, grass mats or even straw may be used for shading. Plastic/tarpaulin sheeting is useful as it also keeps in heat and moisture, which helps to speed up curing. In hot and tropical climates, the curing process can be accelerated by damping down the finished work more frequently, as soon as all moisture has evaporated.

Figure 6.18 Initial curing of blocks – keep shaded and regularly dampened.

6.3.9 Stacking blocks

After the initial month of curing, when they are strong enough to be moved, stack the blocks in the shade. Make sure the ground is level to reduce the risk of stacked blocks breaking. Leave gaps between the blocks to allow air and moisture to circulate (see Figure 6.19). Lime-stabilized earth blocks are best given a minimum of 28 days to harden and cure before use. Hot weather will speed the process and cold weather will slow it. In cold climates it is advisable to avoid winter working, particularly if temperatures fall below 5°C.

FIELD TESTING: STAGE 3 – BUILDING ELEMENTS 139

Figure 6.19 Stacking, curing and storing blocks.

6.3.10 Preparing blocks for final placing – dip in water before use

Dampen well-cured lime-stabilized blocks before use. Just before building, either dip each block into water for ten seconds or lightly spray them where stacked (see Figure 6.20). Do not soak them. If the blocks are too wet or too dry, they will affect the moisture content of the mortar, reducing its bond strength. Keep all lime-stabilized soil mortar work shaded and dampened, both during the work and after placing,

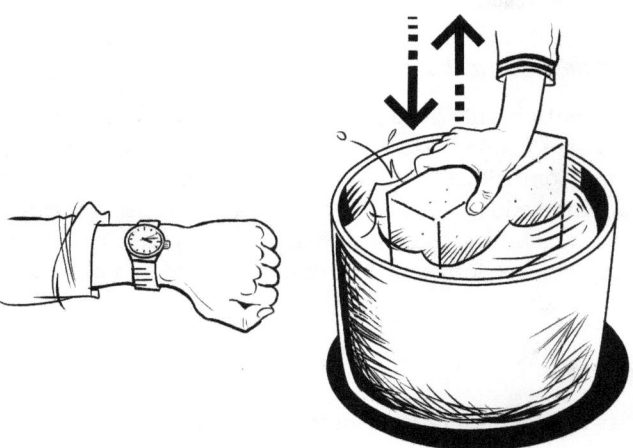

Figure 6.20 Dip the cured blocks in water for ten seconds before use.

6.3.11 Field testing the blocks during the main work

Before starting the main work, plan ahead for the immersion and compressive-strength testing of production-run blocks to check for consistency. Allow for testing at least monthly, and possibly more frequently subject to speed of construction.

Conduct the tests on production blocks one month after curing them, and at regular intervals during production for the main work, or following production of every batch of 500 blocks. Test at least five blocks in each production run. Test as described in Stage 2 field testing (see Section 5.8).

An indication of the rate of compressive-strength development can be given by carrying out simple field tests with a hand-held penetrometer at regular intervals. Test results taken every three months over two years would provide useful information.

Figure 6.21 Keep newly laid block walling dampened and shaded throughout the work to stop the mortar drying out too quickly. This will lengthen its curing time and improve strength.

6.3.12 Machine block-making

Mixes for use with a CINVA ram or other machinery for compressed-earth block-making (see Figure 6.14) need to be a little firmer, and possibly drier, than those used for making blocks by hand. They may not require the addition of fibre and will need to be handled carefully. A two-person team on a single hand-operated machine can usually achieve an average production run of 300 blocks per day. Prepare sufficient flat, shaded ground and protective coverings in advance of production.

6.4 Soil mortar

Lime-stabilized soil mortar (also commonly known as 'earth mortar') will need to have the same durability in wet conditions as the bricks and blocks with which it is used. The same mix as for the blocks can be used for the mortar, although for fine joints a finer aggregate may be preferred with a maximum particle size of about 2 mm (ASTM No. 10). One way to achieve this is to use the same mix as for the blocks, but sieve out some of the larger aggregate particles. Alternatively, use a different soil that has a finer particle-size grading. Do not change the lime–soil ratio for stabilization from that established during Stage 2 testing unless it is for testing a mix in connection with greater workability.

6.4.1 Mortar mix testing

Trial mortar mixes and any variations can be prepared and tested by subjecting 50 mm (2″) cube samples of the mixes to the Stage 2 immersion and wet compressive strength tests. A higher proportion of lime may produce a mortar with improved workability, but careful preparation and immersion testing is needed to ensure that the adjusted mix will remain stable under water. Discs of mortar trial mixes, approximately 75 mm (3″) in diameter and 25 mm (1″) thick, can also be made up for immersion testing after curing, for testing setting times (see Section 6.4.2) and for permeability testing (see Section 5.8.6).

6.4.2 Mortar setting-time tests

Allow discs of the fully prepared mortar sample or a lime-mortar paste, brought to the consistency of bread dough or pottery clay, to harden, cure, and set as recommended in Section 6.3.8, or in ambient UK weather conditions.

The simplest field test during the curing process is to press a finger, held at arm's length, into the mortar at intervals. It is considered set when no depression is created, and there is no alteration to its form until it breaks.

A more controlled setting-time field test may be carried out on similarly prepared mortar samples with a simple piece of equipment. This consists of a hardwood or metal rod that has one end shaped to a square point of 1 mm (1/32″) sides. A standard weight of 300 g (10 oz) is fastened to the other end of the rod. At carefully timed intervals after the moment the sample preparation is completed, the point is lowered to rest on the disc. Initially, the point will continue to sink into the top of the sample. The initial set is taken to be the time between completing preparation of the sample and when the sample bears the point of the rod (sometimes called a needle), without forming a depression in the surface.

6.4.3 Mortar application

If blocks are well made, with square and even sides and sharp arrises, and are of consistent size, they can be laid with fine joints, under 10 mm (3/8″), which require less mortar than for rough blocks. Use a straight edge and a level (or plumb bob) to help build straight. To leave a key for render, set the bedding mortar joints back from the face of the blocks by 10–20 mm (3/8″–3/4″).

6.4.4 Damping and protecting new mortar

If the lime in the mortar dries too quickly, it is likely to crumble and fail. After placing the mortar, keep the walls under shade and dampened for a minimum of one week. During this time, the blocks and mortar need to be moistened to help them cure slowly and continue to strengthen. The longer the walls are dampened and kept in the shade the better: four weeks or more is ideal. If possible, do not build in direct sunlight, and erect a screen to shade all the new work from sun and rain at all times (see Figure 6.22). In cold climates, do not prepare, build, or cure work with lime or lime-stabilized soil when there is a risk of frost.

Figure 6.22 Protect new walls and mortar from drying out too quickly. Keep shaded and regularly dampened.

6.5 Cob

Cob was previously commonly used for house building in the UK, where a quarter of a million cob houses are still occupied today in the south-west of England, with regional variations further east and north. Modern cob construction is being reintroduced in the UK, Ireland, other parts of Europe and in America. This is due to the sustainable construction benefits, standards of finish, and comfortable healthy, living conditions that can be achieved by earth buildings.

Mixes for wall blocks given in Section 6.3 could be a starting point for cob trial mixes, although the mix can be coarser than for blocks. The clay content of soil used for cob should be 15%–25%, with 10%–15% gravel and the rest a mixture of fine and coarse sand with fibre (and possibly cow dung). The best form of lime to add for stabilizing cob is lime putty. The water content of the mix needs to be higher than that for block-making yet should be sufficient to achieve a stiff mouldable consistency. Building with cob is often entirely 'hands-on', so the consistency of the final mix is likely to be best judged as correct when it 'feels' right and performs well. It is also preferable to add more and longer fibres, more clay and coarser aggregate than for block-making, in order to assist handling and the in-situ shaping process, improving workability and bonding. Cob walls are built without formwork on firm foundations, usually of stone or brick, that rise 300–600 mm (1'–2') above ground level and are no wider than the earth wall above (see Figures 6.23 and 6.24). Cob walls can be built with either parallel or tapering sides. In Yemen and Iran, high loadbearing walls have been constructed in this way.

During placing, use of a compaction method similar to that described for foundations of lime-stabilized soil (see Section 6.2.6) is good practice and improves final strength. A common method, after preparing the cob and ensuring the base of the lime-stabilized soil wall is ready to support the mix, is to build the wall, generally 600 mm (2') wide, in 'rises' of about 300 mm to 500 mm high (12" to 20") at a time around the full perimeter of the building, which allows firming up of the earlier sections before returning to build the next rise on them. For traditional cob, each rise is compacted, by pitch fork, spade or through the weight of the builders treading down by foot. Lime-stabilized soil is best compacted well and in thinner layers so the rammed-earth technique described below is likely to be more suitable for solid earth walling in most cases.

For lime-stabilized cob, it is important that the proportions of lime and clay are correct, having been selected following Stage 2 stabilization tests as set out in Chapter 5. Depending on the extent of exposure to wet conditions, it may be that only the more vulnerable areas of cob walling need stabilizing. These might be at plinth level (up to the first 300 mm (1') of walling, or to at least 300 mm (1') above any possible flood level), and adjacent to wet areas such as water supplies and drainage. Un-stabilized cob could be used in less exposed and less vulnerable locations. Publications on un-stabilized cob construction methods include Williams-Ellis et al. (1947) and McCann (1983).

Figure 6.23 Cob wall-making and shaping on solid and stabilized plinth.

Figure 6.24 Cob wall-making and shaping is best if from a raised platform for high walls.

6.5.1 Cob trial mixes

Use the proposed lime-stabilized cob mix to make 150 mm (6″) cube test samples. After curing for 28 days, conduct the immersion test and monitor the performance of the trial mixes over time. Test also for both dry and wet compressive strength at 28 days and then again at two months, six months, and one year in the same way as for other building elements.

Trial mixes should also be assessed for expansion and shrinkage, as mixes that are very clay-rich may crack (particularly if they are not modified by the addition of sand and/or fibre). If quicklime powder is used for stabilization, however, it will expand within the mix, counteracting shrinkage and cracking.

6.6 Rammed earth

Rammed-earth walls are built within formwork (shuttering), directly from the foundations or from a plinth. A prepared mix is placed in the base of the formwork in a layer about 100 mm (4″) deep, and then rammed (highly compressed) with hand rammers or by machine. This is followed by placing and ramming further layers. Apart from the use of formwork, the main difference between rammed earth and cob is the consistency of the mix: rammed-earth mixes are significantly drier than cob mixes. For effective compression, the moisture content of a rammed-earth mix should be between 7% and 15%, which is similar to the minimum required moisture content of a mix for use in block-making. In practice, it means a handful of slightly moist mix can hold its form when squeezed tightly. Powdered quicklime would therefore be the most appropriate form of lime to use for rammed-earth walling, provided adequate maturing time is allowed before placing. In parts of France and Spain where rammed earth has traditionally been widely used as a construction material, it was common practice to ram a lime–soil mix into the corners and along the external face of the wall, within the formwork, to give the walls additional durability.

6.6.1 Rammed-earth trial mixes

Rammed earth usually requires a natural or modified soil with a clay content of between 10% and 20%. Mixes similar to those suggested for machine-compressed wall blocks could be a starting point for rammed-earth trial mixes, although the mix can be coarser. Use the proposed stabilized mix to make 150 mm (6″) cubes and cure and test these in a similar way as for cob mixes, carrying out an immersion test and both dry and wet compressive strength tests after 28 days curing. Subject to the building programme, repeat these tests at intervals of two months, six months, and one year in the same way as for other building elements.

As with cob, the stabilization of rammed-earth walling may only be required in those areas of the building exposed to wet conditions; in dry and well-protected locations, un-stabilized rammed earth would be satisfactory.

The walling is built up in well-compacted layers of 100 mm (4″) similar to the method described above for trench footing foundations, but inside shuttering or formwork. For further information on rammed earth, refer to *Rammed Earth Structures: A Code of Practice* (Keable, 2011).

6.7 Wattle and daub

Wattle and daub constructions are built from a frame of interlaced supports (the wattle) covered on one or both sides with a layer of render (the daub). Generally this is not a recommended construction method for external walls unless the structural framework is strong and the wattle and render are sufficiently robust to withstand extreme weather conditions.

6.7.1 Wattle

Wattle in England is normally composed of lightweight interlaced laths, battens or sticks woven between firm upright staves fixed into a timber frame. The laths may be plastered on both sides provided they are rigid and firmly supported. They need to be evenly spaced and not too far apart to enable a good key for the daub to be formed. It is important that the wattle is tightly woven but with sufficient spaces for the daub to key through, and that there is a strong structural framework on which the daub panels rely for support.

There are many materials suitable for wattle, including grasses and cane. Care is needed to avoid using materials that are not sufficiently strong to support the daub. Thin reeds and grasses, for example, even when interwoven, are unlikely to be sufficiently robust or durable enough to support a wall panel for long.

146 BUILDING WITH LIME-STABILIZED SOIL

6.7.2 Lime-stabilized daub and render

Daub and render may be of similar lime-stabilized soil mixes to those described for render in Section 5.5.2 and Section 6.10, applied to a wattle framework. The mix may be improved for daubing by adjusting and possibly increasing the proportion of clay and fibres. Carry out trials to determine whether a more sticky and slightly wetter mix than for plaster would assist adhesion to the wattle. Daub test samples can be made up individually as discs, and tested the same way as for render and mortar (Sections 5.5.4 and 6.10). Excellent descriptions of various traditional wattle and daub construction methods are given in the SPAB Technical Pamphlet 11 by Kenneth Reid (1989) (See photos 3a and 3b).

Figure 6.25 Typical UK wattle and daub construction.
Source: Based on Reid (1989).

6.8 Internal plaster

6.8.1 Lime for plaster

The term 'earth plaster' is also commonly used for lime-stabilized soil plaster.

Well-burnt, fully reactive, and finely powdered quicklime (particle size less than 0.85 mm/ASTM No. 20), thoroughly mixed with well-tempered and dampened soil, may be used for earth plasters, but well-matured lime putty is safer, and will in many cases be preferable due to the difficulty of obtaining finely powdered quicklime of sufficient quality (particularly so where the quicklime is not stored in fully airtight bags or containers). Well-matured lime putty will need to be of the correct density and have been sieved and allowed to settle out in the settlement pit for at least three weeks – preferably three months or longer.

It is important for good-quality renders, plasters and mortars to screen or sieve out any lumps from the mix during the slaking process. Recommended sieve sizes are given in Table 4.1. Plaster finishing coats are thinner and will require finer aggregate particle sizes and equivalent fine sieves.

Dry hydrate is not recommended for renders or plasters unless it is of the highest quality and fresh from the kiln, as slightly over-burnt particles may be subject to late hydration and may continue slaking in the mix. They may then expand, causing pitting and popping.

6.8.2 Clay-rich soils for plaster

If the selected subsoil layer from the trial hole has a high clay content, compare trial mixes made with quicklime powder with those made with lime putty, proportioned as described in Section 5.4. Similarly, prepare and assess trial plaster panels (Section 6.9). Before mixing, the clay soil should be first dried and sieved to remove any gravel or stones over 5 mm (3/16") in diameter for the first coat, and 1 mm to 2 mm for the finishing coat, then well tempered. Add water until a sticky plaster is made. It should be well tempered for one or two days both before and after adding lime putty, although this would not usually be necessary if adding best quality quicklime powder. However, after adding quicklime to a mix, always check that it has fully slaked before use. Trial different periods of tempering to achieve the best workability with each different soil type.

6.8.3 Sand (or very sandy soil) for plaster

If the subsoil is not clay-rich but very sandy, it may still be possible to use pure lime to make a flood-resilient plaster. There is an extensive history of using all forms of lime with sand only for plasterwork, and the material produced can be of the highest quality. However, pure lime (non-hydraulic lime) does not give a hydraulic set, so pure lime with sand is only suitable for internal plasters in parts of a building that will not be subjected to flooding.

If artificial or natural hydraulic lime is available, it may be used with sand and sandy soils as an alternative to lime-stabilized clay soil where wet conditions are a concern, subject to the full three-stage testing programme. Otherwise, a plaster made from non-hydraulic lime and sand (or sandy soil) that needs to resist long-term wet conditions will require the addition of a reactive pozzolan. All forms of non-hydraulic lime can be used to make such a plaster – crushed quicklime, dry hydrate powder (which, if not available, can be made on site), or lime putty. The active minerals of the pozzolan react with the lime in the mix to stabilize it, compensating for the lack of clay in the soil.

When calculating proportions for trial mixes with pozzolans, refer to Section 5.6.6. Ensure the pozzolan is reactive by first testing it, as set out in Section 4.4.

Initial test mixes (all subject to Stage 1 and 2 testing and plaster panel trial results) for a sandy or very low-clay-content soil for base plaster work, and where fully stabilized-soil plaster is required, could be:

- One part of finely crushed, reactive quicklime powder, or two parts lime putty, plus three parts of successfully tested, finely sieved pozzolan (e.g. finely sieved burnt-brick dust) to four parts or more of sandy soil. Add fibres and just enough water to make a sticky plaster that will not crack when dry.
- One part finely crushed best quality quicklime powder, or two parts lime putty, to two or three parts of finely sieved pozzolan mixed well with 10 to 20 parts low-clay subsoil, one part cow-dung slurry and two (or more) parts short fibres.
- Two parts finely crushed good-quality quicklime powder, or four parts lime putty, and four to six parts of tested, finely sieved pozzolan, six to eight or more parts sandy subsoil, one part cow-dung slurry and two (or more) parts short chopped straw.

Note the variation in mixes above, which gives an indication of the range of mixes that should be tested to determine the best mix for stabilization of a low-clay-content soil using the least amount of lime. In all cases, the constituent parts, including the finely sieved pozzolans, should have satisfied both Stage 1 and immersion testing. A systematic method for trialling the mixes as plaster on the wall is outlined in Section 6.9.

Figure 6.26 When making a plaster mix, initially retain the lime to soil proportions required for stabilization, as described in Section 5.5, with pozzolan if necessary to compensate for a lack of clay content. Modify the mix with the addition of sand and/or fibres to suit local conditions.

6.8.4 Testing and applying internal plaster

The mix for the first (base) coat should always incorporate well-distributed fibres (see Figure 6.27). The finishing coat may have finer, shorter fibre (e.g. hair or finely sieved very short straw) or no such addition, subject to trials. Assess the optimum fibre type and content by testing different types and proportions in sample render panels (see Section 6.9).

Figure 6.27 Add fibre for tensile strength.

Test the plaster mixes to establish their qualities of adhesion and workability by making sample plaster panels in a systematic matrix (example shown in Table 6.2) and, after curing for 28 days, examine for a robust, crack-free and dust-free finish. In addition, and if the plaster is to withstand damp or wet conditions, immerse the sample discs (made of the same mixes used for the plaster samples) in water for a month, to ensure that the same mix is fully stabilized and does not dissolve. Successful mixes could also be tested for water permeability (Section 5.8.6), which is increasingly important for flood-prone areas and wet conditions.

6.9 Render and plaster test panels

Render and plaster wall and ceiling finishes are applied to brick, block, laths and wattle, cob, rammed earth, straw bale and other walling material backgrounds. They improve the durability of the walls, as well as providing a smooth, hygienic, and clean surface finish.

Following test Stages 1 and 2, successful lime-stabilized soil plaster and render mixes should be subject to further testing to evaluate the additional and essential plaster properties of good workability, key, adhesion, and finish quality. (See photo 16b.)

Without changing the lime to soil proportions for stabilization as established in Stage 2 testing, prepare trial render/plaster mixes with increasing proportions of fibre, sand (if available), and possibly a pozzolan (see Table 6.2). Apply these as they are to be used in the finished work as trial panels onto a prepared, keyed background. The trial panels could be as small as 250 mm by 250 mm (about 10″ × 10″), but in order to give an indication of the best performance and quality that can be achieved, 600 mm by 600 mm (about 2′ × 2′) is preferable. If space constraints dictate a test matrix of the smaller size panels, consider making 1 m × 1 m (about 3′ × 3′) test panels of the three most successful results from the matrix, subjecting them to a more robust examination and selecting the best performance. The preparation of these test panels also provides experience in assessing their adhesion and workability as well as the techniques of applying new materials and mixes.

The panels should be prepared as outlined in Section 6.10.1 and applied to a dust-free, well-keyed and wetted surface. It is best to apply the mixes manually using a float, trowel, or by gloved hands rather than with a spray machine. Compaction during application is important, as it improves durability. This may be achieved by trowelling on hard and scouring with a plasterer's float, throwing on (from a 'harling' or 'roughcast' trowel), or throwing on vigorously by hand and smoothing and scouring down afterwards with a float or trowel. These three different ways may be used separately or in any combination depending on local custom.

Ensure all mixes are applied in the same manner, and all samples have the same dimensions and thickness. Carefully scratch details of the mix proportions towards the top or bottom edge of each panel before it dries and when 'leather hard'. Keep all panels well shaded and cure for 28 days. Some initial results may be apparent within two or three days.

The panels can be used to establish the optimum mix for:

- good adhesion of the first coat with the mix well keyed and firmly applied to the background;
- a first coat that will firm up to provide a solid substrate well keyed on its surface to take the subsequent and finishing coats;
- ease of application;
- minimum cracking/shrinkage;
- a finishing coat which is well bonded to the previous coat;
- a fine, dust-free, robust surface to the finish coat that does not crack or scratch easily – important for all finishes, particularly external render which may also have a textured surface;
- a clean, flat, polished, and attractive finishing coat for internal plaster.

Select the mixes that perform the best in line with the above properties and which have proved successful in the immersion tests. Mixes that have been modified or revised for specific building elements should be re-tested as for Stage 2 immersion testing.

Table 6.2 Render Panel Matrix: Example of render or plaster base-coat test panels – each of different trial mixes with increasing amounts of fibre (from left to right) and sand (from top to bottom). (See also photo 16a.)

Increasing fibre parts ⟶

		0	1	2	3
Increasing sand parts ↓	0	Base Mix Only eg 1LP:6So*	1LP:6So:1F	1LP:6So:2F	1LP:6So:3F
	1	ILP:6So:1Sa	ILP:6So:1Sa:1F	ILP:6So:1Sa:2F	ILP:6So:1Sa:3F
	2	ILP:6So:2Sa	ILP:6So:2Sa:1F	ILP:6So:2Sa:2F	ILP:6So:2Sa:3F
	3	ILP:6So:3Sa	ILP:6So:3Sa:1F	ILP:6So:3Sa:2F	ILP:6So:3Sa:3F

Base Mix: The base mix should be based on successful stage 1 and 2 trial mixes of lime to soil proportions. (In this example, the LSS base mix is 1LP:6So.)
*no fibre or additional sand parts are added to the base mix in the first panel
Key: LP (Lime putty); So (Soil); Sa (Sand); F (Fibre)
Note: For simple field-testing purposes, these parts can be measured by volume.

An easy way to test for thinner finishing coats of render, plaster, or limewash trial mixes is to key (scratch) half of each selected trial panel of the base coat to use as a substrate onto which to apply and test the finish coat. A base coat may be in the region of 10–20 mm thick (⅜"–¾") and a finish coat is generally only a few millimetres thick, either lime-rich or made with fine aggregate. In all cases where there are different mixes for various plaster coats, each mix needs to pass all three field-test stages.

It may be that a base coat will need to be applied onto an earlier levelling coat. For levelling and base coats, apply a coarser and more fibrous mix for crack-free and strong bonding properties. A sticky, fibre-rich mix can be applied more thickly without slumping, and can level out undulations in the wall face. A fibre-rich mix may also add a degree of thermal insulation. Any fine cracks that occur in the base coat after it has cured can be filled with the finish coat when it is applied. If the mix has been successfully tested as a trial panel, larger cracks should not occur.

Several other methods of eliminating larger cracks include using more compaction; pressing back before the mix has hardened; reducing the amount of fine aggregate; adding more fibre and/or sand; reducing the moisture content; and ensuring a level substrate (avoiding differences in thickness of the render/plaster, as different thicknesses dry at different rates and so introduce the likelihood of cracking).

Thin finish coats may require fine sharp sand and/or other fine stone dust or pozzolan to add a degree of robustness and create a harder surface, particularly for external render. If trial mixes for finish coats include fibre, the fibres should be significantly finer than those in the backing coat, sieved as fine as possible, and well (evenly) distributed in the mix.

It may well be practical, and save time, to prepare and cure discs of the various trial render mixes at the same time as the trial panels. Cured trial

152 BUILDING WITH LIME-STABILIZED SOIL

discs made with the same mix as the most successful (robust, crack-free, and well-bonded) trial panels can then be subject to immersion testing for stability under water as detailed in Stage 2, and to permeability testing. The optimum mix will be successful in all tests.

If a low-clay soil, or sandy soil, or sand forms the base for trial external renders, then mixes in the region of one part lime putty to two or three parts sand and one or two parts pozzolan could be trialled, with added fibre in proportions from one through to three parts, and subjected to panel tests and Stage 2 immersion testing. Sharp and well-graded sand will give best results but, depending on availability and cost, tests can be conducted with both soft and sharp sand, individually or with a mixture of the two. An excellent way to learn and develop skills in the design and use of mixes for plasters and renders with local materials is to test progressive series of trial mixes, setting them out in a matrix as shown in Table 6.2 (and photo 16a) to cure, document (label, date/photograph), and record the results.

6.10 External render

Suggestions for test mixes for initial trials of external render mixes are given in Section 6.10.3. External walls and plinths may be rendered with a similar mix to that used for plaster, but it could be made more hardwearing by using coarser (and sharper) aggregate (Table 6.2). (See photos 16a, 16b and 16c.)

6.10.1 Preparation

Background keying. Renders need a solid, clean and absorbent background. A good key (see Figures 6.28 and 6.29) is important and may be provided by:

- leaving the bedding mortar between blocks set back from the face;
- raking out the joints;
- scratching or roughening the whole surface;
- building the wall with an absorbent and rough surface in the first instance.

Figure 6.28 Preparing the wall by raking out the joints.

FIELD TESTING: STAGE 3 – BUILDING ELEMENTS   153

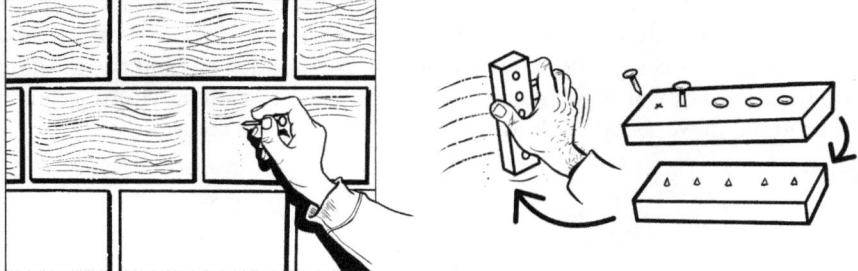

Figure 6.29 Preparing the wall by scratching a 'key' into the surface.

Once keying has been completed, the wall should be brushed down to remove loose dust and debris and then damped down, as renders will not stick to very dusty surfaces.

Figure 6.30 Brush down to remove all loose debris and dust.

Prepare weather protection and sun shading. Prepare protection material (sheeting/tarpaulins, reed mats, sacking, large cloths, or sheets) for shading the walls of the building from direct sun and heavy rain (see Figure 6.31).

This is very important because the lime in the render mix needs to be kept moist (but not too wet) while it cures, hardens, and continues to carbonate over several weeks. Following compaction, carbonation, pozzolanic activity and chemical set will be assisted by warm, moist conditions. Adequate strength and initial set may be achieved in weeks, whilst long-term strength gain may continue for years.

Figure 6.31 Prepare shading material in advance.

Give the greatest protection to walls that are exposed to direct sun or to the most rain. Attach sheeting or other protection firmly to the eaves or at the top of newly rendered surfaces, to ensure it will stay in place for a month. Keep the temporary protection well away from the walls so that the render is not damaged and can be damped down regularly.

Figure 6.32 Provide shade for all lime-stabilized work.

Damping down. Wet the walls thoroughly at least an hour before rendering (Figure 6.33). Then damp them lightly again just before rendering. Render will not adhere to dry, dusty surfaces or surfaces that are too wet. If the walls are not damped down, they are likely to take moisture out of the render too quickly, which will reduce its adhesion, and lime-stabilized soil mixes must not dry out too quickly or they risk failure.

Figure 6.33 Wet the walls before applying render.

Lime quality. Renders that incorporate best quality, finely powdered quicklime produce a fast and strong reaction but need to be applied within 20 minutes before they stiffen and set. Mix in small quantities, making sure to combine materials well, and apply immediately.

Well slaked lime putty is a safer choice, with good workability, for plasters and renders than quicklime, unless strict regulation and quality control of quicklime production and distribution are in place. Some mixes made using fresh lime putty may be improved by maturing for one or two days, provided they are kept damp. Include test samples with different maturing times in the Stage 3 test programme before the main work.

Storage and protection of mixes. Use all mixes before they start to set or dry out and harden.

Mixtures that incorporate (best quality) powdered quicklime should not be stored and should be used as soon as possible following mixing, once all the quicklime has fully slaked.

Generally, all mixes using non-hydraulic hydrated lime should be kept damp and used quickly, or within a day or two of mixing. During the course of the work, store prepared lime putty-based mixes in the shade and cover with damp cloths or sheeting to keep them from drying out (see Figures 6.9 and 6.34).

156 BUILDING WITH LIME-STABILIZED SOIL

Figure 6.34 Storage and protection of mixes.

6.10.2 Render application

As far as possible, avoid extreme hot or cold weather conditions when applying render. Prepare a trial area before starting the main work. (See photo 16b.)

Base coat. Use wooden or steel floats, or, if these are not available, gloved hands to apply a layer of render no more than 20 mm (¾") thick to dust-free and dampened walls (see Figure 6.35). The optimum thickness will depend on the render materials available, quality of the background, and number of coats of render planned. When applying by hand, adhesion can be improved by throwing the first coat of render hard onto the wall's surface.

Figure 6.35 Apply render firmly to dampened background – if applying by hand, wear gloves.

Closing cracks. If two or more coats of render are to be applied, a small amount of cracking in the first is seldom detrimental. The ideal mix will show little or no cracking, and any that may appear could possibly be squashed closed, but the tendency of the first render coat to crack can be reduced in a number of ways. These are similar to the methods for preventing plaster from cracking, listed above, and include:

- keeping the water content of the mix low;
- compacting the render well onto the wall;
- using best quality, finely powdered quicklime instead of lime putty or dry hydrate;
- incorporating plenty of fibres in the mix;
- ensuring a strong key to the background;
- ensuring good aftercare by shading and damping down regularly;
- incorporating cow-dung slurry;
- reducing the proportion of the finest aggregate.

Figure 6.36 Closing cracks.

Second and subsequent coats. If finishing or/and levelling coats are to be applied, scratch-key the previous coat before it fully hardens, when it is leather hard (green-hard), and wet the walls again immediately before applying the new coat.

Safety precautions. Protect eyes when plastering, particularly when working above head height. Keep eyewash and plenty of clean water ready for flushing out any material that is splashed into the eyes.

158 BUILDING WITH LIME-STABILIZED SOIL

Figure 6.37 Damp down each render coat before applying the following coat.

Curing. As previously described, shade all render from hot sun and rain and cure it for 28 days. Keep protective sheeting clear of the wall surface so it does not damage the finished face. Cure by lightly spraying or flicking water with a wetted brush. Wet the render as frequently as possible throughout the first two weeks; then, for the final two weeks, wet the render a minimum of three times a day, more if in high temperatures. Keep the shading curtain damp. A bucket of water underneath it will ensure water for damping is always readily available and will help keep the air humid near the wall face (see Figures 6.21 and 6.37). The more cycles of wetting and drying out, the quicker the render is likely to carbonate and durability improve.

6.10.3 Render trial mixes for additional durability

The durability of lime-stabilized soil renders can be improved in various ways, some of which are described below.

Using the correct proportions of lime to soil. This largely depends on the type and quantity of clay. Either too little or too much lime in the mix may reduce the durability of the render. Finely powdered, best quality quicklime is the most reactive stabilizer, but only perfectly burnt quicklime should be used for render (as late slaking can spoil the finish). Subject to production methods, lime putty can be very effective and may be easier to apply as it helps workability and creates a naturally sticky mix, although it is slower acting and is required in different proportions to quicklime. For best results, ensure only well-burnt and fully reactive powdered fresh quicklime or well-prepared lime putty is used for the soil stabilization of finishes, particularly renders, plasters, and screeds. Methods of preparing and testing lime to soil proportions are given in Chapter 5.

Selection of aggregate. It is important to use hard, sharp, well-graded aggregate of appropriate size for different render coat thicknesses. The aggregate may either be contained in the soil or added for modification when mixing. It might be clean sand and/or crushed inert material such as sandstone, limestone, chalk, or marble. Coral ragstone falls into this category. Finely crushed, even powdered, calcium carbonate from any source helps to seed calcite growth and slowly strengthen render. If it is a hard material, such as marble, and not too finely crushed, it will also add strength.

Addition of pozzolan. This may be beneficial if there is insufficient active clay in the soil for stabilization with lime, and/or if additional compressive strength or impact resistance is required. Some lime–pozzolan only mixes will produce screeds that are strong and close to impermeable. (These are not always suitable as render mixes where vapour permeability is required.)

Addition of fibres. Adding fibres of suitable length and quantity to the mix improves binding qualities, increases tensile strength, and reduces the risk of cracking. Suitable organic fibres include straw, jute, hemp, and hair. It is advisable to add chopped fibres approximately 25–50 mm (1"–2") long to all base coat mixes. However, straw and readily biodegradable fibres should be avoided for areas subject to flooding if an inert fibre is available. Fibres are particularly important additives if the lime stabilizer is in the form of putty or dry hydrate, although best quality, powdered quicklime may itself be sufficient to eliminate shrinkage or cracking. As a starting point for the quantity of fibres to be added to render, Cowper (1998 [1927]) notes that the traditional amount was half a pound of hair to one cubic foot of mix, which is approximately 0.25 kg to 0.3 m^3 (½ lb to 1 cubic foot or one bucket). Ensure the fibres are evenly distributed throughout the mix.

Organic additives. Plant and animal oil derivatives applied to finished surfaces improve water-shedding properties but are seldom durable. Adding small quantities to the last finishing coat of limewash or render may have a similar effect but last longer. A suitable amount is around 3% of the lime proportion of the mix; adding more oil than this tends to reduce permeability and binding qualities. *Earth Construction* (Houben and Guillaud, 1989) lists numerous plant and animal extracts that have been used as water repellents in conjunction with soil, or for additional stabilization. The effectiveness of these materials is variable. Some are documented in publications about historic building fabric but many are traditional and used locally in different parts of the world. They include linseed and other oils, animal fat/tallow, cow dung, rye or high gluten wheat flour boiled in water to produce a paste, sugar, egg, various natural forms of soap, ashes, beeswax, and blood.

Under certain conditions, particularly in damp environments, organic additives may promote mould growth. This can be prevented by the introduction of a mould inhibitor to the mix, although where surfaces dry out quickly, this may not be necessary. Boiling oil and other liquid additives before adding them to the mix may also reduce the risk of mould growth. Lime itself has inherent antiseptic properties and naturally inhibits mould growth.

Additional durability. Consider variations to initial mixes if additional durability is required for elements such as window sills and copings, and for areas involving water supply, storage, and drainage. Mixes could include one or more of the following additional materials: pozzolan; short chopped fibre or hair; well-graded sharp sand; coarse sand and fine gravel; crushed or ground kiln or blast-furnace slag; very hard marble or limestone dust or grit; slurried cow dung. Add only enough water to make a mix plastic and workable. Panel testing (see Section 6.9) may show that some mixes are satisfactory without adding or increasing the proportion of these ingredients; this will depend on the soil's mineralogy.

Render reinforcement at joins, corners, or junctions. A layer of render may need to be applied across different background surfaces: for example, at junctions between wall panels and structural elements, or where there are changes in walling material. In these cases, there are a number of ways to provide additional strength over the join, including floating damp hessian (open-weave) sacking or other fibrous material into the render at the junctions. Another vulnerable position where such reinforcement could reduce the risk of cracking is at lintel bearing ends above the top corners of window and door openings; in this case, float the fabric at a 45° angle (see Figure 6.38).

Figure 6.38 Possible locations for render reinforcement with additional fibres to reduce risk of cracking.

Test mixes for lime-stabilized earth renders with clay content. Prepare both trial render panels and disc samples of render mixes to test. Select final mixes that immersion tests have shown to be fully stabilized, and that panels show perform well in terms of strong adhesion to the background, minimum cracking, and robustness of finish. Where greater impermeability or impact resistance is required, trial sample mixes containing increased proportions of finely sieved pozzolan. Impact resistance could also be improved by adding a proportion of marble or limestone grit instead of sand.

Trial render with clay-rich soil. The lime to soil proportions for render are to be determined by Stage 2 testing as set out in Section 5.4.2.

An example of a modified mix to test for lime-stabilized soil render could be two parts lime putty to between 10 and 20 parts well-tempered clay soil plus three to five or more parts fibre (hair or short chopped straw) and/or one part cow-dung slurry (where this is traditional practice). Use a shovel or paddle mixer to turn the mix over until it is thoroughly mixed and has a consistent colour. Using quicklime rather than putty (or different locally available sands) can create useful variations in trial mixes as described for trial plaster mixes in Section 6.8. Test each mix over a range of tempering periods from half an hour to two days and trial the quicklime mixes to check they are fully slaked.

Trial render for low-clay soils. Maintain the lime to soil ratio required for stabilization determined through Stage 2 testing as above. If there is little or no clay in the soil, trial mixes with added pozzolan. To make a lime-stabilized render where the subsoil has a low clay content and a high sand content, an appropriate trial mix might include two to four parts lime putty, four to six parts fine brick dust, and 12 to 18 or more parts of sandy soil, preferably sharp and well graded (see Figure 6.39). All such mixes should only be used subject to Stage 2 and Stage 3 testing.

Figure 6.39 Test trial render mixes containing a variety of materials including quicklime or lime putty, and different sands and pozzolans. Trial different proportions of cow dung and fibre. Initially maintain lime–clay–pozzolan proportions determined by Stage 2 testing.

The properties of lime-stabilized soil mixes can be varied by the addition of modifying agents, typically pozzolan, fibres, or cow dung. These may enhance the performance of the building element for which the mix is produced by improving water resistance, tensile strength, and plasticity respectively.

Where using cow dung is common practice, consider testing mixes with an additional half to one part slurried cow dung and two or more parts of short chopped fibres. Make sure the same size containers are used to measure all materials. Only add enough water to make the mix sufficiently plastic for workability. Test and adjust all final mix proportions to ensure they pass the immersion (soak) test before use. (See Section 5.8.2 and photos 15a, 15b and 17c.)

6.11 Screeds and impervious surfaces

Suggestions for floor screeds and pit lining trial mixes are given in Section 6.11.3.

6.11.1 Floor finishes

In many areas of the world there is a long tradition of using un-stabilized floors, made by combining well-tempered soil with chopped straw and cow dung. These floors can perform satisfactorily under soft shoes, bare feet or carpet, provided that the components have been thoroughly mixed, compacted, and dried and remain dry. Un-stabilized mixes, however, are water soluble and therefore deteriorate rapidly when wet. There are a number of ways to improve a floor's wearing qualities and its resistance to damp and wet conditions.

A firm, hardwearing, dry, water-resilient and insulated floor is usually the general objective. A common approach to achieving this at ground level is to build the floor up in a series of layers using the same principles as for three-coat plastering but on a larger scale. Thus, the first base layer is the thickest and contains the largest aggregate, and each successive layer is thinner and contains finer aggregate than the one below. Hardness and impermeability throughout can usually be improved by increasing the proportion of very fine pozzolan in the mix (see Table 4.1), as well as ensuring thorough compaction of each layer. The finishing layer may contain hardening and polishing agents to add strength, reduce the formation of dust from the surface, and improve appearance.

Hard wearing materials that can be laid into floor surface finishes include small flat stones, bricks or tiles, used whole or as broken pieces set in a surface screed to form a mosaic. Larger pieces could be set on a floor screed using strong hydraulic lime and sand, or lime and pozzolan mixes for the bedding and pointing mortar.

There are many alternative ways of building up floors with earth and each needs to be appropriate for the local climate and site conditions. Useful references to a wide variety of earth floor construction methods are set out by Houben and Guillaud (1989) and Historic England (2015).

FIELD TESTING: STAGE 3 – BUILDING ELEMENTS 163

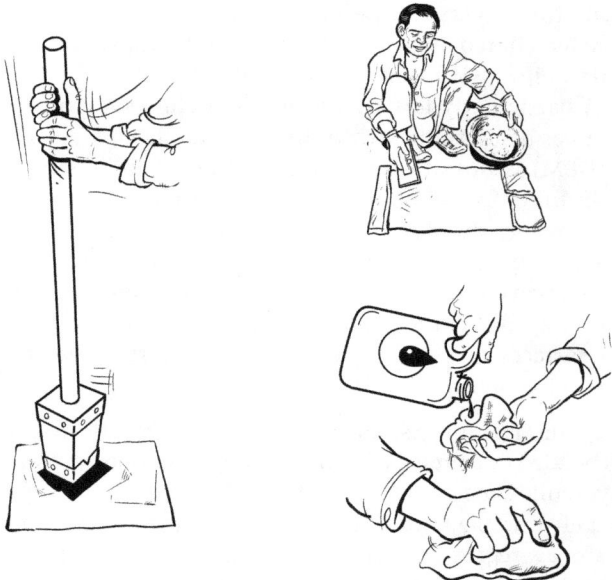

Figure 6.40 Compress and compact floor screeds well, trowel surfaces flat and level, and consider oiling and polishing for additional durability.

The following sequence describes one way in which a floor could be built up:

1. Soil sub-base levelled and compacted.
2. Well-compacted hardcore of broken stone or similar material, such as broken brick or large gravel with particle size 25 mm to 75 mm (1" to 3") as a first layer, to a total thickness of 150 mm to 225 mm (6" to 9").
3. Compacted lime-stabilized soil with well-graded aggregate of fine gravel and coarse sand (see Table 5.1) in a layer 100 mm to 150 mm thick (4" to 6"). The surface of this is scratch-keyed when it is leather hard (green-hard).
4. Compacted lime-stabilized soil incorporating well-graded fine sand, in a layer 50 mm (2") thick, mechanically well bonded to the previous layer with a scratch key, and laid and compacted when the previous layer is leather hard.
5. The floor should be cured before final surface treatment and polishing. The procedure for curing and aftercare is as described previously for all mixes incorporating lime.

Additional layers of other materials may be added between the hardcore and lime-stabilized soil layer as appropriate to the design for local conditions. This may include the addition of insulation and/or the reduction of dampness by creation of a capillary break with appropriate materials.

Thermal insulation layers can be of inert materials, such as 150 mm (6") thick, compacted charcoal, or, in industrial areas, proprietary products such as vermiculite bound with lime, or recycled foam-glass (RFG) or expanded blown clay. A particularly low-cost example includes laying empty bottles (containing air as insulation) within the lime-stabilized layer. A damp-proof membrane (DPM) is not always necessary, but if one is required, options include tar, bitumen or rubber solutions, polythene sheet, or a clay layer. Whatever is used, it should be applied to a lime-stabilized soil blinding (surface covering) over the compacted hardcore layer.

The surface finish depends on several factors including laying skill, purpose of the room, cost, appearance, and materials available. Water-resistant and hardwearing surfaces will be needed in particular for areas such as WCs, bathrooms, and kitchens.

- The most durable finishes are well-laid hard brick, tile, or polished stone pavers bedded in a lime–pozzolan mortar, accurately laid with fine joints and hydraulic lime grout.
- Moderately durable finishes may be based on a mixture of strong binder and hard aggregate or surface inclusions. If softer aggregate, such as broken brick and marble chippings or soapstone, is used, the floor surface can be ground down to give a smooth flat finish, and polished to create an attractive terrazzo. Creating polished terrazzo or a mosaic is labour-intensive and requires skill, but if laid with care, both are attractive and durable finishes.
- Softer finishes which may be comfortable to walk on but may not withstand heavy impact have the advantage that they are probably the least costly and least demanding to prepare. Subject to local conditions and testing, a development of the traditional soil, chopped straw, and cow-dung mix with an appropriate addition of lime for stabilization may be useful. Mixes of lime and pozzolan, and sand or lime, soil, pozzolan, and hard aggregate are also worth testing for screeds. To prevent dusting, finished surfaces will need oiling or waxing and polishing after they have cured and dried.

A finish detailed by Weismann and Bryce (2006) uses a cob mix of clay subsoil sieved through a 3 mm ($^1/_8$") mesh, mixed with short chopped fine straw and sand. This mix would need to be tested in the same way as described for renders above. A layer 12 mm to 25 mm (½" to 1") thick is applied to a clean and dampened sub-floor layer (of lime-stabilized soil), compacted down and trowelled to a level and well-keyed surface. When it is leather hard (green), a final layer 12 mm (½") thick is applied and polished smooth by trowel or float. Once dried, it is sealed, often with boiled linseed oil, but if this is not available other plant oils could be tested. (Mustard seed oil, for example, is more common in Nepal.) The oil is often used in conjunction with a turpentine or citrus thinner in four applications: the first coat is 100% oil, the second is a mix that is 25% thinner, the third is equal parts oil and thinner, and the last 25% oil and 75% thinner. Each coat is allowed to dry before applying the next. The finished surface is then polished with beeswax, which has been melted to liquid and mixed with two parts oil. In addition to testing locally available oils and thinners, it is advisable to apply the full three-stage testing programme to earth-based layers, both with and without the addition of lime for stabilization, before starting the main work, as these will enable a comparative evaluation of the different materials and mixes. All earth floor layers that are subject to flooding and wet conditions need to be stabilized. There are many alternative ways to build floors with earth and these need to be appropriate for local climate and site conditions. Useful references on various earth floor construction methods are set out in Houben and Guillaud (1989) and Historic England (2015).

6.11.2 Water storage and pit linings

In order to resist water or contain it, screeds or renders should be impermeable or made of a mix that could be described as very hydraulic. These are necessary for lining tanks or pits to hold water and for floor screeds or wet areas in a house. Mixes that are likely to achieve this will need to include high proportions of pozzolan, or clay and non-hydraulic lime, or hydraulic lime. One part non-hydraulic lime to three parts very fine pozzolan, with one or two parts of fine sharp sand (or eminently hydraulic lime and sharp sand), is typical of mixes that may be required (see Figure 6.39 and 6.40).

6.11.3 Trial mixes for screeds and pit linings

Some trial mixes for hardwearing screeds, water retention and lime-slaking pits are suggested below, but due to the variable nature of materials, particularly soils, samples must be fully tested and confirmed satisfactory before use in the main work.

Additional protection against water damage could be considered. Possibilities are the inclusion of a small amount (2% to 3%) of water repellents such as linseed or other oil, slurried cow dung, or tallow in the finish. A number of authorities state that potassium sulphate (K_2SO_4), also referred to as sulphate of potash and used in gardening, will help to 'waterproof' soil (see, for example, Houben and Guillaud, 1989). But each additive considered needs to be tested to determine whether it offers any benefits over lime stabilization alone.

The wearing qualities of a screed may be improved by adding marble dust, grit, crushed limestone aggregate or other ground, hard materials to the mix. These could be used to replace the sand.

Suggested mixes for testing, subject to the availability of materials (including short fibres, which, if used, would be omitted from the finishing coat), include the following:

- 1 quicklime powder : 10 finely sieved clay-rich soil : 2 sand and/or pozzolan, e.g. ground and finely sieved brick dust (the linear shrinkage test (see Section 5.3) will establish best guidelines for how many parts of clay-content soil to one part crushed quicklime for hydraulic set)

166 BUILDING WITH LIME-STABILIZED SOIL

- 1 lime putty : 3 pozzolan : 1 sharp sand
- 1 quicklime powder : 2 pozzolan : 2 sand
- 1 quicklime powder : 3 pozzolan
- 1 quicklime powder : 2 pozzolan : 3 crushed limestone or marble grit
- 1 hydraulic lime : 2 sand and/or grit (see Sections 6.3, 6.4 and 4.1.8 for a description of hydraulic limes).

In any of the above, the addition to or substitution of sand with powdered and sieved limestone, or marble dust and/or grit, or crushed limestone or other hard, inert material, may achieve a harder-wearing finish.

The field testing of samples that have been well prepared and cured for 28 days is the best way to make a comparative evaluation of the different mixes.

6.11.4 Protection and aftercare for screeds and pit linings

Protection of mix. Keep the mix covered and protected from hot sun and rain (see Figure 6.41). If using a hydraulic lime or a powdered quicklime mix, use immediately after the lime has been well mixed in, for best results and strength.

Keep humid. Keep the work area shaded and a source of water close by for regular damping down, and to keep the air near the floor surface humid, if possible.

Curing. Moisten the screed by lightly spraying or flicking the finish with water several times a day for a minimum of one week, but preferably four weeks, following application. In tropical climates, keep sacking protection or other covering moist, to extend evaporation time. This will help to strengthen the screed against wear and reduce permeability.

Figure 6.41 Cure lime-stabilized soil floor slabs and screeds by keeping them shaded and lightly dampened for one to four weeks before final surface finishes or treatments such as oiling and polishing.

Testing. The performance of mixes will vary depending on subsoil type and local materials used, so it is important to field test all these, as previously described for plaster and render. Make trial samples and test after at least one month's curing to ensure the proposed mix is satisfactory and fit for purpose.

6.12 Finishes for lime-stabilized soil backgrounds

6.12.1 Decoration

Various forms of relief or raised decoration may be integrated with the render face on rendered walls, using the same mix as the render itself, or a modification of it. Consider adding additional hair (or short fibre), fine sharp sand, limestone dust, or marble dust to mixes for testing. These materials may help workability and strengthen delicate decoration against abrasion and rainwater. Scratch and wet the wall to provide a mechanical key before applying additional material to ensure a good bond (see Figure 6.42). In England, this type of traditional decoration is called pargeting (See photo 5b).

Figure 6.42 Scratch-key the render and moisten it before applying raised decoration.

6.12.2 Limewash as a paint

After curing plasters, render and pargeting decoration, if any, and while the shading remains in place, both external and internal wall surfaces may be finished with three or more coats of limewash. This will help seal any fine cracks and protect the plaster from water damage. A pure lime or slightly hydraulic lime may be used as the principal binder for limewash.

6.12.3 Limewash preparation, application, and aftercare

Wall preparation. A lime-stabilized soil surface, whether un-plastered, plastered or rendered, is an ideal background on which to apply limewash.

168 BUILDING WITH LIME-STABILIZED SOIL

After wall surfaces have been cured, and before preparing limewash, ensure sheeting material is in place ready to shade the work. For best adherence, first brush down the walls to remove dust and loose particles (see Figure 6.43). Thoroughly damp down the face of the wall about half an hour before application, and again a few minutes before starting, to ensure that the thin coat of limewash bonds well with the wall and does not dry out too quickly (see Figure 6.44). Limewash will need the same damp and slow drying conditions for carbonation as other lime-stabilized elements. Ensure, however, that the walls are not too wet, nor running with surface water, before limewash is applied.

Figure 6.43 Brush off the walls to remove dust and debris before damping down and applying limewash.

Figure 6.44 Damp down the walls before limewash application.

Mixing limewash for external use. The limewash may be of water and lime putty from pure or slightly hydraulic lime, or a mix of well-diluted lime putty with very fine pozzolan (approximately one to two parts pozzolan to six parts fresh putty, depending on pozzolan reactivity test results). A good limewash can also be made by slaking and sieving quicklime directly, provided a very fine sieve is used (see Section 4.1.2 and Table 4.1). It is important that the quicklime has been tested to check that it is the best quality.

Add sufficient water to bring the limewash to the consistency of thin milk. Mix it thoroughly and pass it through a fine sieve (0.18 mm or ASTM No. 80; see Figure 6.45). For best results, limewash should be very thin: it is better to apply as many as five very thin coats, all well absorbed into the background, rather than one or two thick coats that would be likely to crack and peel or flake off the wall. Ensure that the limewash is well stirred, and stir again at regular intervals during application to keep the lime particles in suspension all the time.

Figure 6.45 To make limewash, add enough water to best quality lime putty or quicklime to bring it to the consistency of thin milk and pass it through a fine sieve.

Application. When limewashing, it is important to protect your eyes. Wear goggles. Work at a cool time of day or in mild weather conditions. Do not apply limewash or lime render during frosty weather or when temperatures may be below 5°C. With a stiff natural bristle brush, apply thin coats to dampened walls using a hard scrubbing action (Figure 6.46). Allow 24 hours between coats. Dampen the limewash down again several times after it has dried: as for all work with lime binders, the speed and extent of carbonation will be increased if the damping down process is repeated more frequently. Keep the walls protected from direct sun by shading at all times of day until the final coat has cured. In cold climates, it is also important to protect each coat against unexpected night-time freezing conditions until it is fully cured and dried.

Figure 6.46 Apply limewash to wetted and shaded walls – not in direct sunlight.

Second and subsequent coats. After 24 hours or more, including curing, apply a second coat of limewash to re-wetted walls. Repeat the process to build up a suggested minimum of five coats, applying each no sooner than 24 hours after the previous one. The last one or two coats of limewash may have a small amount of earth pigment and/or oil added (see below).

Figure 6.47 Cure each coat of limewash for at least 24 hours before applying the following coat.

Pigments. If colouring is required, add small amounts of natural earth pigments to the last one or two coats. Generally, it is recommended that the quantity of pigment is kept below 6% of the lime putty to maintain durability (see Figure 6.48).

To maintain a consistent colour, the quantity and quality of all materials including water need to be carefully recorded, and the proportions of all materials, particularly pigment, should be the same for each batch. It is a good idea to mix enough to cover an entire wall at one time. Additionally, store a tightly covered bucket of the pigmented limewash for future use to 'touch up' if and where necessary, as making a batch to replicate the original shade can be difficult.

Figure 6.48 If colour is required, add small amounts of earth pigments and mix well.

Additives. Improved water-shedding properties have traditionally been achieved by adding a small proportion (approximately 3%) of organic extracts/oils, such as coconut, linseed or a similar oil, to the last finishing coat of limewash. (Note that if an oil additive is applied prior to the last coat of limewash, it will reduce the adhesion of the following coat.)

Additions to the last coat of limewash may be broadly categorized under the headings of oils, fats, glues, gums, and resins. Readily accessible additives of this nature vary with local conditions but those with a long history of use in the area are often appropriate.

If the option is available, first test the addition of a small quantity of oil. Adding only 2% to 6% of oil to the mix for the final coat can be a surprisingly effective way of helping limewash shed water, and it is a simple finish to apply.

Vegetable oils suitable for field testing include raw linseed oil, coconut oil, mustard seed oil, castor oil, shea oil, cotton oil and various nut oils. Animal fats and oils that could be used include tallow, lard, fish oils, skimmed milk or casein. Soft soap could also be tested. If more conventional oils are not available, it may be worth carrying out trials with sap and oils extracted from regionally available sources by crushing and boiling selected parts of plants such as mimosa, baobab, and acacia.

Protein-based additives such as casein can support mould growth, however, so are best avoided in damp environments unless boiled before use or used with mould inhibitors or disinfectants such as formalin, flowers of sulphur, copper sulphate, carbolic acid, or coal tar.

Fine aggregates of very fine sand or stone dust may also be added in small quantities to improve general durability.

Additives to limewash are best incorporated while it is hot, immediately after slaking or during slaking from the quicklime. They may also be added to cold limewash prepared from lime putty or dry hydrate, but the effect may not be as durable.

If various forms of lime and local additives are available, comparative field testing of finished samples is a good way to establish the optimum mix – and may also be an opportunity to gain useful experience of using local (and possibly new) materials and mixes before starting the main work. A method of field testing limewash formulations and paints is by coating a render test disc specimen with the proposed fully coated and cured finish, and then subjecting it to a permeability test as described in Section 5.8.6 and illustrated in Figure 5.18.

A sample panel coated with a selected number of fully cured coats of the proposed limewash finish can be tested through a series of simple, mainly visual tests over time, as described in *Building with Lime* (Holmes and Wingate, 2002). Inspect the finish for adhesion, cohesiveness, colour-fastness, weathering resistance, splash resistance, mould resistance, salts resistance, surface quality, and permeability. Additionally, the limewashed samples can be subjected to a 'hose pipe' test to monitor the performance of the proposed finish, whereby a fixed spray of water is directed at the sample for a fixed period of time to replicate likely extreme weather conditions on the exposed face of the building.

Well-prepared and well-applied limewash can remain sound without attention for ten years or more. This is to be expected under moderate weather conditions, but well-prepared and well-applied limewash has also been known to withstand climatic extremes.

Final curing, aftercare and protection. During and following application, keep limewash fully shaded from sun and protected from rain. Lightly damp the surface down several times a day after it has dried. Continue this process for a week or more for best results (see Figure 6.49). Temperatures have a marked effect on set and curing times. High temperatures, which could be up to 40°C (104°F) or more in some parts of the world, help speed the set if it is kept regularly dampened, whereas low temperatures of, say, 5°C (41°F) or below could delay set and carbonation for months. The best working temperatures, which also minimize the length of time non-hydraulic limes must be treated in this way to ensure that they are well cured, are usually between 10°C and 30°C (50°F and 86°F).

Figure 6.49 Fully cure solid lime-stabilized soil backgrounds, such as blocks and render, by keeping them shaded and dampened for up to 28 days, before applying the limewash. Keep each coat of limewash shaded, and cure it for a minimum of 24 hours before applying the next coat.

6.13 Roof finishes

Finishes for flat roofs may be tested in a similar way to floor finishes and screeds (see Section 6.11.1).

However, roof finishes and roof screeds have different design requirements to ground-floor slabs. Supported by the superstructure, roof finishes and screeds should have some degree of flexibility, unlike ground-floor slabs and floor screeds when constructed directly off the ground. Suspended (upper) floors and roofs need to have a more lightweight construction and so timber boarding is appropriate and often used. Alternatively, roof screeds may benefit from the rapid drying of a hot-mix application of a lime-stabilized soil material, thereby avoiding the loading of a wetter mix and longer drying time, although this will still require structural support.

Flat roofs are not always appropriate for buildings built from lightweight natural materials such as timber frames or wattle and daub. Where they are used, they should be designed to shed water as quickly as possible and discharge it clear of the walls below.

174 BUILDING WITH LIME-STABILIZED SOIL

For traditional low-pitched roofs, the supporting structure must be firm and sufficiently robust to ensure the screed and finish do not flex so much under their own weight (which will be substantial) that they give rise to subsequent water ingress. However, a degree of flexibility is desirable in a roof screed, as it will minimize the chance of cracking, so trialling mixes with additional fibres or cow dung is recommended. Screeds of this nature can protect against water ingress provided they are well designed and placed, properly tested, and there are good falls and arrangements for the discharge of rainwater. Even so, they may be more suitable for use as a substrate to take tile or other hardwearing finishes. As with other lime-stabilized building elements, ensure all stages of roof finish are well shaded and cured for 28 days.

For pitched roofs, precast or fired-clay tiles offer a more durable finish than screed on matting or earth, which is often used to create sloping roofs with a low pitch in the tropics. Well-laid, steeply pitched thick thatch is an excellent low-cost, environmentally benign and effective solution, provided good thatchers are available. Thatch is likely to offer more comfort and be cooler for the occupants than a flat or low-pitch roof as it is a better insulator than most other readily available roofing materials.

Figure 6.50 Thatch, well laid to a sufficient depth at a good pitch, can be an excellent roofing option, especially in terms of weather resistance, thermal efficiency, cost, and environmental benefits.

6.14 Three-stage field testing – conclusion

The careful selection and field testing of low-cost, local materials and lime-stabilized soil mixes, can support the construction of durable buildings that will remain stable in wet and flood conditions for many years.

That a wide range of soil mixes across large and differing geographical areas can be successfully stabilized using lime, is clear from the thousands of lime-stabilized soil houses built this way.

FIELD TESTING: STAGE 3 – BUILDING ELEMENTS   175

Due to the variability of soils and clays, in situations where laboratory analysis is not possible, it is important that those wishing to do the same follow the three stages of field-testing procedures described in Chapters 4 to 6 and summarized in Figure 6.51 for consistent and reliable results.

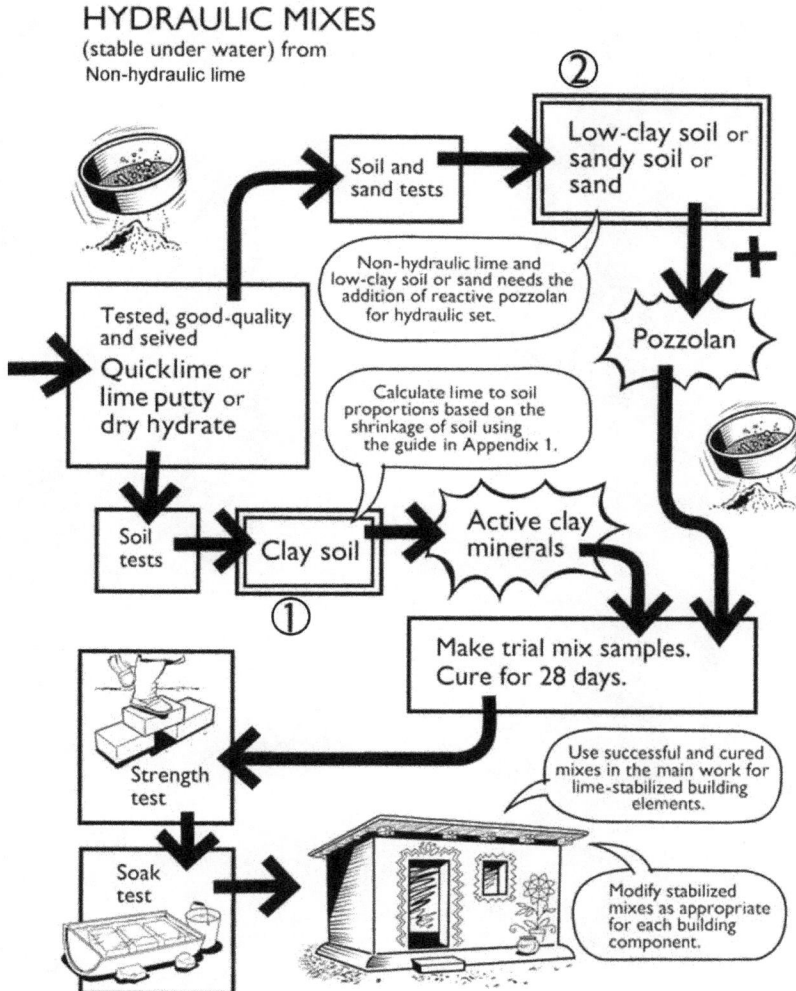

Figure 6.51 Testing hydraulic mixes for lime-stabilized soil – summary. (See photo 11 for the colour version.)

Once the three-stage field-test procedure has been completed and lime to soil proportions and mix ratios have been established (for a building site or village), few further trials should be necessary. However, it is advisable to monitor and test mixes for quality during component manufacture and throughout the construction programme.

Photo 1a Old Fort, Zanzibar: also known as the Arab Fort (Swahili: Ngome Kongwe): a 17th-century fortification built by the Omani Arabs, with lime-stabilized soil mortar and render to the massive ragstone walls. The fort is the oldest building in Stone Town, the capital of Zanzibar. (Sections: 2.2 & 2.8)

Photo 1b Fort Jesus Mombasa: (Portuguese: Fort Jesus de Mombaça): a 16th-century fort located on Mombasa Island. Although a fine example of western renaissance architecture, the masonry techniques including the lime-stabilized soil mortar and render used for the entire structure and ramparts, are believed to have been provided by the local Swahili people – techniques still in use for residential construction centuries later. (Sections: 2.2 & 2.8)

Photo 1c Alhambra Palace and fortress complex (the complete name in Arabic means 'red fortress'): constructed in the mid-14th century by the Moorish rulers of Granada on a plateau overlooking the city, with defensive walls and towers of compacted lime-stabilized soil. (Sections: 2.2 & 2.8)

Photos 1a,b,c Three examples of the use of lime-stabilized soil in the construction of major fortifications, all of which are UNESCO World Heritage protected sites: (Sections: 1.2 & 2.2, 2.3, 2.8)

Photo 2 Rear view of lime-stabilized soil plaster showing nibs keyed to laths. Principal staircase wall of a Grade I listed 17th-century manor house, UK. (Section: 2.7)

Photo 3a 16th-century hall house, Warwickshire, UK. Timber frame with lime-stabilized soil wattle and daub external panels and lime-stabilized soil internal wall and ceiling plaster. Repaired with like-for-like matching materials. (Sections: 2.5 & 6.7.2)

Photo 3b Detail from 3a, of lime-stabilized soil wattle and daub panel before repair. (Sections: 2.5 & 6.7.2)

Photo 4a Lime-stabilized soil plaster mix with 5% powered quicklime prepared for immediate use, UK. (Section: 2.7)

Photo 4b Lime-stabilized soil ceiling plaster undergoing repair in the house shown at 3a, with like-for-like material. Master Plasterer Jeff Orton leading Trumpers' plastering team, 2001. (Section: 2.7)

Photo 5a Lime-stabilized lateritic soil mortar, plaster, and render to ragstone walls, ceilings and floor slabs in the 19th-century Ithnashery Dispensary, Zanzibar. (Sections: 2.3 & 2.8)

Photo 5b Detail of weathered lime-stabilized lateritic soil render and decorative mouldings to the building shown at 5a, completed c.1900. Photographed in 1989 before repair. (Sections: 2.8 & 6.12)

Photo 6 Lime-stabilized soil mortar to stone wall of historic farm building c.1850, National Trust Holnicote Estate, Somerset, UK. (Section: 2.6)

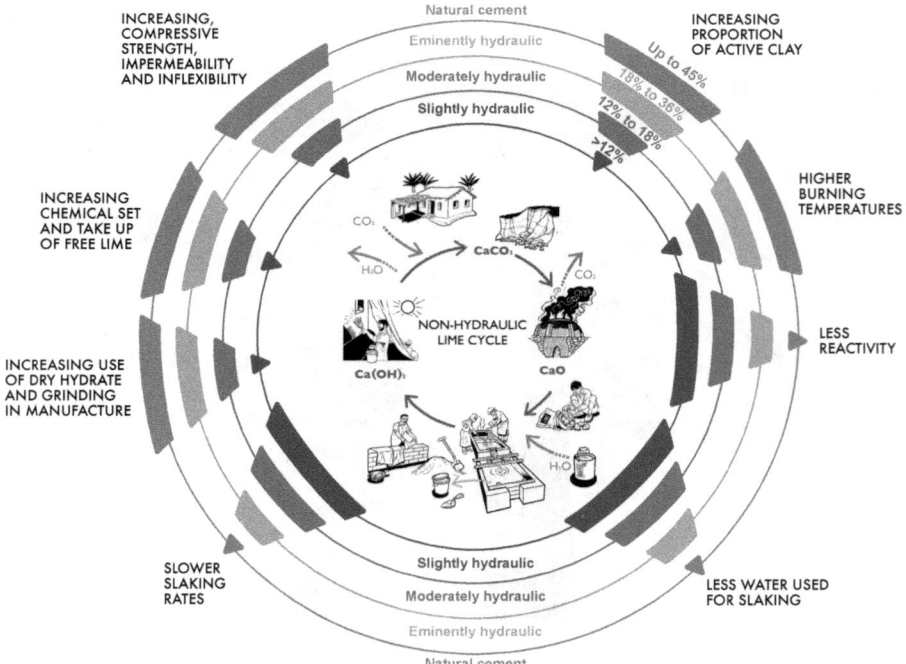

Photo 7 The lime cycle and related properties of hydraulic limes. (Section: 3.3)

Photo 8 Typical small-scale lime burning and production, Pakistan, 2015. (Section: 4.1.4)

Photo 9 Lime-stabilized soil field kiln, burning local limestone at BLF (Building Limes Forum) Gathering, Cirencester, UK. (Section: 4.1.4)

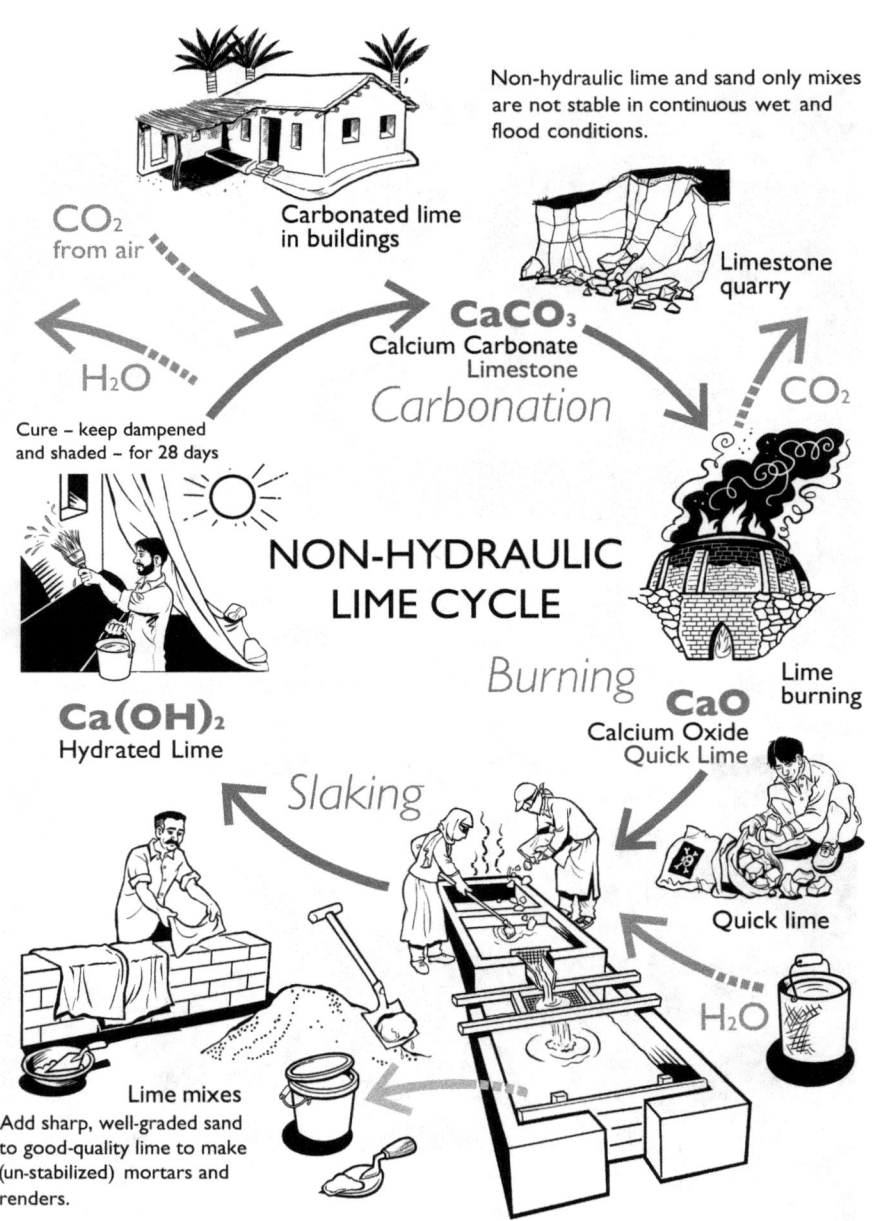

Photo 10 The non-hydraulic lime cycle.

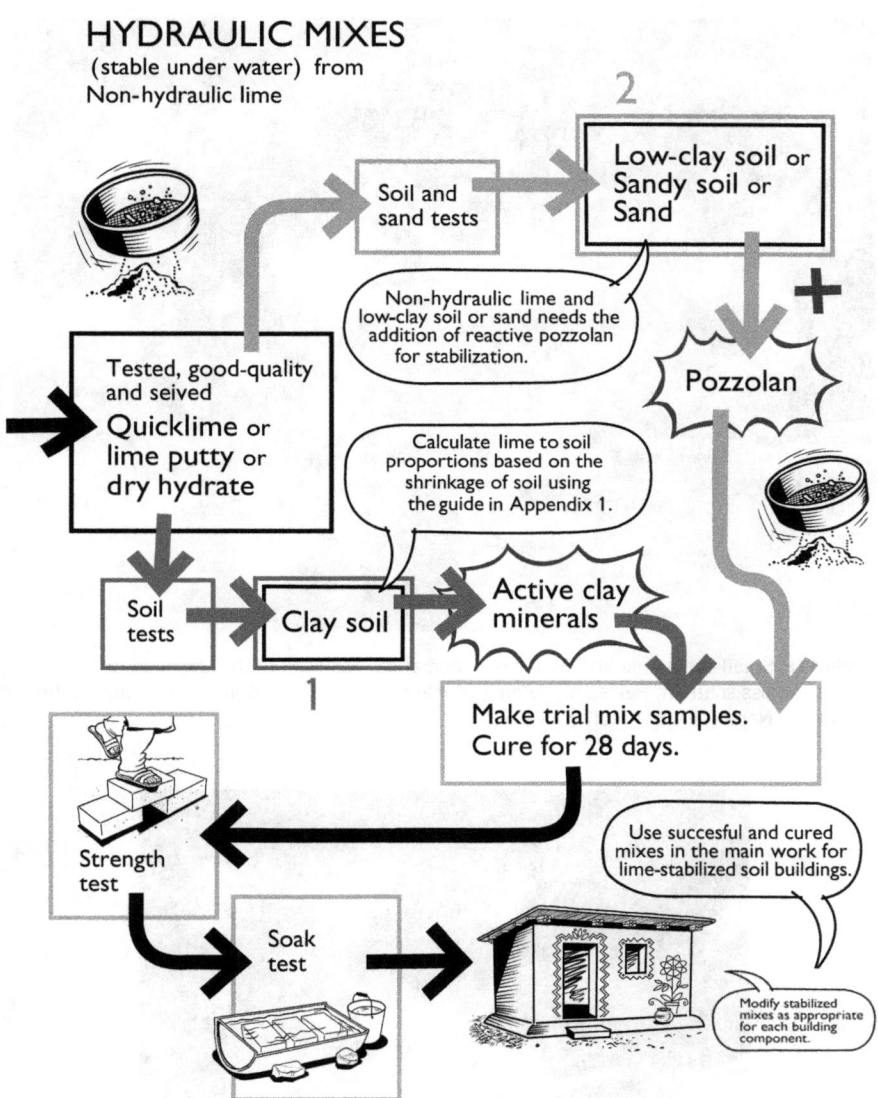

Photo 11 The process of obtaining and mixing materials to produce a hydraulic set from non-hydraulic lime with (1) clay content soil, or (2) low-clay content soil, sand or sandy soil (plus pozzolan).

Photo 12a Small-scale lime-slaking tanks in operation – as used in thousands of rural villages across southern Pakistan. Training course at ABARI Earth Building Training Centre, Dhulikhel, Nepal, 2016. (Section: 4.1.9)

Photo 12b Lime-slaking tanks cut from recycled oil drum, in operation for field testing and as lime-slaking demonstration on training course with KRMEF, Kathmandu, Nepal, 2018. (Section: 4.1.9)

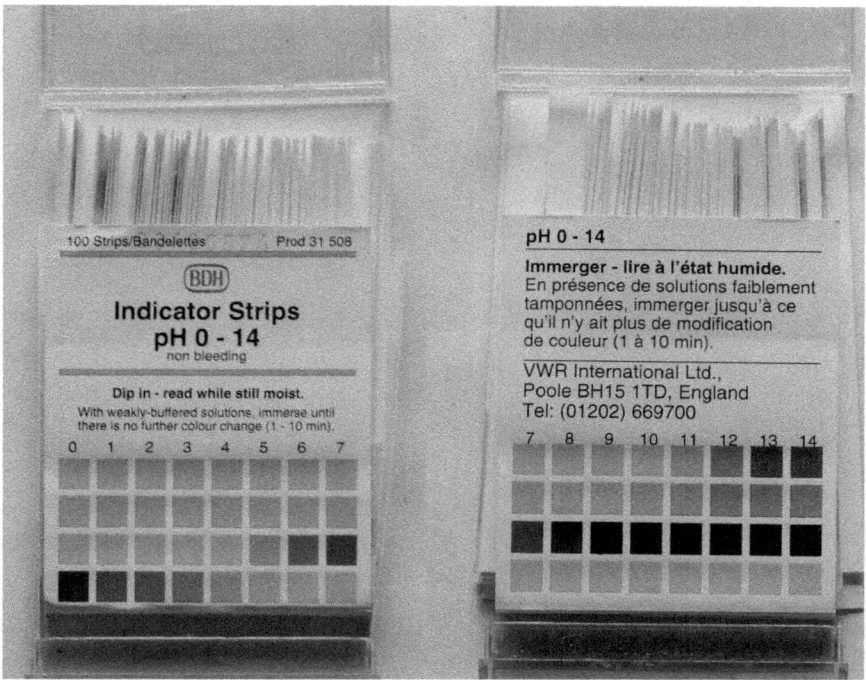

Photo 13 Colour reagent indicator strips for assessing soil pH. (Section: 5.4.4)

Photo 14 Compacted-earth block-making machine in operation, Nepal, 2016. (Section: 6.3)

Photo 15a Soak test: Lime-stabilized soil block (1 part of lime putty : 10 parts of soil) after 12 months continuous immersion in water to confirm lime:soil proportions for stabilization. nNGO HANDS Pakistan, 2015. (Section: 5.8.2)

Photo 15b Wet-compressive-strength testing of the same block after one year's soaking, at a penetrometer maximum reading of 4.75 N/mm² (700 psi), Pakistan, 2015. (Section: 5.8.5)

Photo 16a Render Panel Matrix: Example of a series of render or plaster base coat trial mixes with increasing parts of fibre (from left to right) and sand (from top to bottom). LSS training course at ABARI Earth Building Training Centre, Nepal, 2016. (Section: 6.9, Table 6.2)

Photo 16b Lime-stabilized soil render test panel, being applied to STCDA wall in Zanzibar. Stone Town Conservation and Development Authority, Zanzibar. (Sections: 6.8, 6.9 & 6.10.2)

Photo 16c Applying earth–lime render repair mix to an 18th-century Kathmandu palace external wall following damage from the 2015 earthquake. (Sections: 2.8 & 6.10)

Photo 17a Trial-mix samples. A selection of trial lime-stabilized soil mixes prepared in a range of moulds ready for 28-day curing: cubes, discs and cylinders in a variety of sizes for different building elements. LSS training course for Scott Wilson Engineering, Nepal, 2018. (Section: 6.4.1)

Photo 17b A small selection of lime-stabilized soil mixes from varying locations in Pakistan, successfully tested in the field in 2015 and selected as examples for laboratory testing and analysis after training courses for nNGO HANDS, UN Agency IOM and iNGO ACTED. (Sections: 3.9, 6.4.1 & 8.2)

Photo 17c Immersion testing of cured lime-stabilized soil trial block samples in the field, Nepal. (Note: water immersion tanks could be made in the same way as the slaking tanks, to avoid the use of plastic.) (Sections: 5.8.3, 5.9 & 6.3.11)

Photo 18 Lime stabilization of M25 motorway sub-grade, London, UK, c.1985. (Section: 7.1)

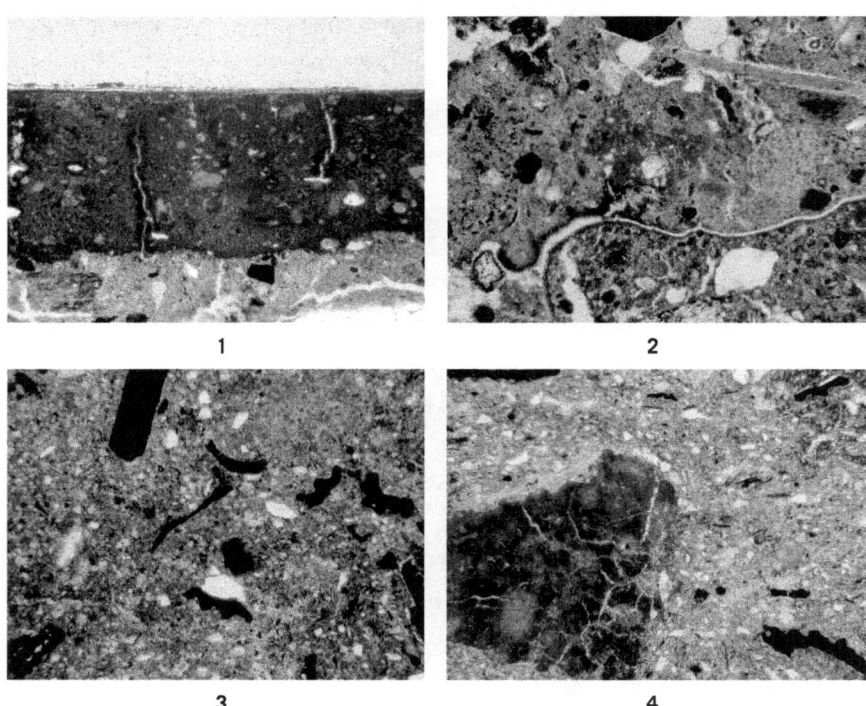

Photo 19 Plates 1 and 2 are magnified thin sections through lime-stabilized soil plaster at a Grade I listed 17th-century English manor house, examined for English Heritage in 2016. Plates 3 and 4 are magnified thin sections of lime-stabilized soil mortar samples produced in Pakistan during the 2013–2015 flood-relief programme. All images were taken using planepolarized light and the bluedyed impregnation resin used to prepare the samples highlights the porosity of the matrix in each case. Field of view 2.7 mm. *Source*: All plates produced by CMC Ltd. (Sections: 2.1, 3.9 & 8.2)

CHAPTER 7
Lime-stabilized soil in civil engineering

7.1 Early 20th-century applications

In the mid-20th century, initial attempts were made to revive the lost art of constructing roads with lime–soil mixtures, similar to ancient techniques used by the Romans. These initial scientific trials failed due to a lack of technical knowledge at that time. Subsequent research and advances in laboratory analysis improved understanding of lime–soil chemistry, and this helped to achieve reliable results.

Pioneers of using lime-stabilized soil for road building were members of the Texas Highway Department, who built experimental roads in 1945. The USA's National Lime Association (NLA) gave early research and development support to organizations in the USA that wanted to use this method of road construction.

Little changed during the following decade, but by the late 1950s the benefits of treating soil with lime were more widely understood. Its application for road construction increased rapidly across the United States and there was a growing interest in the UK. In 1953 the first edition of the British standard, entitled *Methods of Test for Stabilized Soils (BS 1924)*, was published.

Road-building programmes in the USA in the late 1950s accelerated. This, combined with progress in lime-stabilized soil technology by the NLA, resulted in a major increase in projects using lime stabilization. In America alone, during the 12 years up to 1969, the amount of lime-stabilized soil used grew from about 100,000 tons to 900,000 tons in one year. By 1979 its use had spread to many industrialized countries, principally for building roads and airport runways. By 1977 knowledge of its use and benefits, by those specializing in road construction, was spreading worldwide. Engineering and geological reports at the time confirmed that most lateritic soils, of which there are widespread deposits in Africa, are well suited for stabilization with lime. They record, for example, that numerous quarries bear evidence of the economic importance of laterites as a road-building material in many areas of eastern Nigeria (Madu, 1977). Also, in connection with limestone deposits in Nigeria, there is a major potential for the use of hydrated lime in place of cement for soil stabilization. These articles also record that in many of the East African countries a high proportion of the bitumen-surfaced roads are constructed on lateritic soil stabilized with hydrated lime (Ola, 1977).

178 BUILDING WITH LIME-STABILIZED SOIL

By the 1970s, the aggregate supply for road building in the UK had become scarce and its cost was increasing. Further development work was undertaken and continued into the 1980s, mainly in connection with airport developments in the southeast of England. In 1986, the Department of Transport *Specification for Highway Works* included details of a method for lime stabilization of subgrades (road base course and substructure) (Department of Transport Highways Agency, 1986). Part of the M25 motorway around London was constructed using this technique. (See photo 18.)

Lime-stabilized soil private estate road completed prior to surface dressing, Warwick, c. 1986.

7.2 Research and development

Research and development of the structural properties of lime-stabilized soils and aggregates have been steadily developing for the last 60 years. Dallas N. Little has carried out an extensive evaluation of research findings for the NLA (Little, 2000). This work has been a valuable source of information and support that has helped to clarify the relevance and benefits of recent research into lime-stabilized construction methods described in this book. Volume 1 is an 89-page summary of his findings and associated research. From these, some of the significant stages in the understanding of the way lime-stabilized soil performs are set out in chronological order below. These research findings, added to insights from traditional historic construction field testing for both conservation and building anew, support the principles of preparation, application and aftercare described in the preceding chapters.

7.2.1 Plasticity reduction (1960)

Eades and Grim, reporting their research in the American *Highway Research Board Bulletin* 262, wrote that information from a worldwide database confirmed that some very plastic soils (with a plasticity index of over 50) can be rendered non-plastic with lime. The reaction depends on the soil's mineralogy, but almost all plastic soils show a plasticity reduction.

7.2.2 Exposure to water (1962)

In the Road Research Laboratory Technical Paper entitled *Investigations to Assess the Potentialities of Lime for Soil Stabilization in the United Kingdom*, Dumbleton confirmed results of investigations into the durability of lime-stabilized soil which concluded that prolonged exposure of lime-treated soils to water has only a slightly detrimental effect and the ratio of soaked to un-soaked unconfined compressive strength is high. The saturation level of lime-stabilized soils is seldom over 90%.

7.2.3 Pozzolanic strength gain (1962)

Eades, Nichols and Grim, writing in *Highway Research Board Bulletin* 335, reported on their evaluation of three different Virginia soils. The soaked California bearing ratio (CBR – see glossary) for each showed a cured strength increase from less than 5% to near 100%. X-ray diffraction and scanning electron microscopy verified that pozzolanic material was responsible for the strength gain in these particular soils.

7.2.4 Autogenous healing (1969)

In *Highway Research Record* 263, Thompson and Dempsey confirmed that lime–soil mixtures possess the ability to heal when conditions are right. If a lime–soil mixture has adequate lime to ensure long-term reactivity, it can recover from minor fracture damage (in this research, due to frost) because of subsequent pozzolanic reactivity, subject to appropriate (warm) ambient temperatures. Ordinary Portland cement (OPC) cannot recover in this way.

7.2.5 Canals (1975–1985)

The effect of lime under hydraulic conditions has been described in *Chemistry and Technology of Lime and Limestone* and there is an excellent example of this illustrated by the Friant-Kern irrigation canal in California (Boynton, 1980). This followed the report by Gutschick (1978), writing in *Pit and Quarry*, who confirmed visual observations of the canal, which was of lime-stabilized soil construction. He reported that after 25 years in use, it maintained slope integrity and resistance to erosion, even though it had been subjected over that time to extreme water flow conditions. These included times of high flow as well as dry periods.

7.2.6 Long-term strength gain (1977)

Kelley, in his article entitled 'A long range durability study of lime-stabilized bases at military posts in the Southwest', *NLA Bulletin* 328, reports investigation of lime-stabilization performance on five US military bases, including Fort Chaffee, Arkansas, where material had been placed in 1949. The investigation showed that the compressive strength of the lime-stabilized base at Fort Chaffee had reached approximately 12.8 N/mm^2 (over 1,880 psi) when tested over 25 years later.

7.2.7 Effect of time and temperature on strength (1978)

An evaluation of lime stabilization strength gains for 12 different soils in California carried out by Doty and M.L. Alexander showed that a seven-day cure at 38°C was roughly equivalent to a 28-day cure at 23°C. All soils gained substantial strength increases between 180 and 360 days, and the strength of some improved to more than 10 N/mm^2 after curing for one year.

7.2.8 Optimum mellowing period (1996)

For research in connection with the Department of Transport's *Specification for Highway Works* (Department of Transport Highways Agency, 1986), Holt and Freer-Hewish investigated mellowing periods of lime-stabilized soil. British Standard test methods were used to determine the unconfined compressive strength of these mellowed for various lengths of time. This resulted in 12- to 24-hour mellowing periods being recommended, before placing the mix.

7.2.9 Critical lime proportions (1997)

Uddin et al., cited by Dallas N. Little (2000) for the NLA, tested lime–clay mixtures with different lime proportions ranging between 2.5% and 15%, and subjected the mixes to 180 days curing. Maximum strength gains were achieved with 10% lime but decreased with 15% lime. When the mix had the optimum lime content, the strength was approximately 11 N/mm^2 (1,600 psi), representing an 11-fold strength gain.

7.3 Industrialized processes, plant and equipment

Industrial processes used for large-scale construction of lime-stabilized road sub-bases and base courses are set out at length in *NLA Bulletin* 326 (ARBA, 2004). The principal operations employ automated transport equipment, plant, and machinery. Whilst this work is on a far larger scale than for most applications envisaged here, the principal processes of preparation, application and aftercare remain near identical.

7.3.1 Transport and delivery

Dry hydrate and quicklime are transported in self-unloading tanker trucks. For small projects, quicklime is sometimes delivered in well-sealed dumpers. Alternatively, lime slurry can be prepared on site in portable slaking tanks, which have a 20–25 ton quicklime capacity, or at a more remote central plant before delivery. Machinery has been developed for site mixing of slurry without storage tankers. It operates on the basis of continuously charging water at a pressure of 4.9 kg/cm^2 (70 psi) with lime dry hydrate at a 65:35 weight ratio in a jet-mixing bowl. This produces lime slurry for immediate pumping into spreader tankers.

7.3.2 Scarification

Scarification of the ground to be stabilized is carried out by mechanized purpose-designed grader scarifiers or disc harrows, operated from the driver's cab, through which non-soil materials (mostly stone) larger than 75 mm (3") are removed. Scarified surfaces have the advantage of holding the lime in place before mixing, particularly in poor weather conditions. Alternatively, following advances in machinery design for deep-mixing lime and soil in situ, the lime is now often applied without scarification.

7.3.3 Lime application

Mechanical means of evenly spreading lime in each of its three main forms – quicklime, dry hydrate, and slurry – have been developed.

Granular quicklime can be pneumatically distributed from self-unloading trailers or trucks with built-in spreaders. Grain sizes vary from 6 mm (ASTM sieve No. 3) for pneumatic spreading to 12 mm (½") for mechanical spreading with augers. An alternative way of spreading quicklime of a larger size, up to 18 mm (¾"), is by bottom-drop trailers followed by spreading with a motor grader.

Dry hydrate powder can be efficiently spread by self-unloading bulk tanker transport trucks. It should not be spread during windy conditions or in populated areas. The use of quicklime granules or slurry rather than powder reduces the risk of dust.

Slurry is usually applied to a scarified surface by distributor trucks using gravity or pressure spray bars to spread the slurry. The truck tanks should incorporate recirculation equipment to keep slurry in suspension. The amount of available lime in the slurry is measured by a standard specific-gravity test.

7.3.4 Mixing and mellowing

A rotary mixer with a water truck attached can be used for mechanical mixing and watering of lime and soil. The soil moisture content is brought to 3% above optimum, then, following mixing to the depth specified (between 150 mm and 450 mm), the layer is shaped and lightly compacted with a sheepsfoot or pneumatic roller. The initial mellowing period depends on the soil type but can be between one and seven days. After this time, it is remixed for final compaction.

7.3.5 Final compaction

Sufficient mechanical mixing and pulverization with a rotary mixer is required to ensure that all material (excluding stone) passes a 25 mm (1″) sieve and 60% a 5 mm ($^3/_{16}$″) sieve. Final base compaction is completed with heavy pneumatic rollers, vibratory padfoot rollers, or other roller combinations. Final surface compaction is normally done with a steel wheel roller.

7.3.6 Final curing

Before a subsequent sub-base or layer is placed, the first layer is cured by either moist curing or membrane curing. Moist curing is by light sprinkling and rolling to maintain the surface in a moist condition. For membrane curing, the top of the compacted layer is sealed with bitumen or a similar emulsion.

7.4 Technical guidance

A number of helpful guidance notes and model specifications for roads and other civil engineering projects using lime-stabilized soil which have been and continue to be produced are summarized below. National standards which support these, or are closely related to the use of lime-stabilized soil, are described in the following chapter.

7.4.1 Department of Transport

In 1976 the UK Department of Transport issued a guidance note covering the lime treatment of soils. This was followed by a full highway works specification in 1986 (Department of Transport Highways Agency, 1986) which was amended in 2001 to incorporate specifications for a capping layer of soil stabilized using lime only, lime and cement, or cement only. The later British and European Standard BS EN 14227:2006 (parts 10 to 14) was in turn superseded by BS EN 14227-15:2015, *Hydraulically Bound Mixtures – Specifications – Hydraulically stabilized soils,* published in January 2016.

LIME-STABILIZED SOIL IN CIVIL ENGINEERING 183

7.4.2 British Lime Association

The British Lime Association (BLA) has produced model specifications for lime-stabilized soil road base layers. BLA technical data includes results from research on the inclusion of ground granulated blast-furnace slag (GGBS) in lime–soil mixes. This confirms that soils with a sulphate level four times higher than that normally permitted for use with lime alone can be successfully stabilized when GGBS is incorporated in the mix.

The BLA provides further technical information which is valuable guidance on preparing and using lime-stabilized soil and pozzolans. The following selected items from the BLA specification are relevant to previous chapters.

Road bases. For road base layers 95% of soil with hydraulic binder (SHB) (pozzolan) should pass a 63 mm (about 2½") sieve. Generally, organic matter should not exceed 2% by weight. Soil containing sulphates must be tested for volume stability in accordance with BS 1924. The curing period of the test specimen should be 28 days, and this should be followed by an immersion period of 32 days. At the end of this time, the immersed strength of treated cohesive soils should be no less than 65% of their non-immersed strength. The strength of treated granular soils at day 60 should be 80% of the non-immersed strength. Generally, tests for quality control are carried out for every 500 m^2 of material at least three times each day.

Lime characteristics. Lime is to be as quicklime with a CaO (calcium oxide) plus MgO (magnesium oxide) content of not less than 80%, and an MgO (magnesium oxide) content of not more than 10%. In accordance with BS EN 459-2, the reactivity with water should be at least 60°C (140°F) within 25 minutes. The particle size of quicklime granules is to be such that 100% pass a 10 mm (½") sieve and 95% pass a 5 mm (ASTM No. 4) sieve.

High-clay soils. Where lime is used to break down soils with high clay content, it should be added and mixed between one and three days before the addition of pozzolan, GGBS, pulverized fuel ash (PFA), or hydraulic road binder (HRB). Moisture content should not be less than the optimum moisture content (OMC) in accordance with the 2.5 kg Proctor test method described in BS 1924:

> *The Moisture Condition Value (MCV) test is a very quick and effective way of determining moisture content in a fine-grained soil. It is particularly useful for site testing. It may be used in laboratory testing by preparing samples at values of 14, 11 and 8. A value of 12/13 normally relates to OMC and 8 to the wettest condition compatible with satisfactory placement, compaction and trafficking.*

OMC can also be determined in accordance with ASTM D 698 or BS EN 14227-15:2015.

Final strength. The compressive strength of treated and compacted soil should not be less than 1 N/mm^2 (147 psi). This is considered sufficiently robust for site traffic prior to constructing subsequent road layers. At least 1.5 N/mm² (220 psi) is recommended where heavy trafficking is expected. An alternative to the Proctor compressive-strength test method is the soaked CBR test referred to in Section 7.2.3. A CBR value of 100% may be considered equivalent to 1 N/mm² (147 psi) compressive strength.

7.5 Examples of completed projects

7.5.1 Airport runways

In the USA, the Federal Aviation Administration (FAA) has produced information and specifications for the construction of runways using soil treatment methods including Advisory Circular AC 150/5370-10A, Part 2, Item P-155 Lime-treated subgrade (no date). Runways constructed using lime for soil stabilization, include those at Newark, Denver International, and Dallas Fort Worth airports. Specifications for airport runways have been regularly up-dated and the current (2018) FAA Advisory Circular describing this method of construction using stabilized soil is in AC 150/5370-10H.

7.5.2 Dams and flood mitigation

Lime-stabilized soil to an appropriate specification and method of construction can provide adequate hydrophobic resistance for earth. It is possible to lime stabilize the whole core of a dam but a more cost-effective method is to use an outer lime-stabilized soil covering about 300 mm to 400 mm (12″ to 16″) thick, although thorough compaction on steep slopes is difficult.

US engineers carried out remedial earthworks in California and following severe flooding on the Mississippi River in 1973. Lime was incorporated into salvaged expansive clay from failed levees and transported to site where it was built up lift by lift to form a lime-stabilized soil earth dam (Boynton, 1980). Similar earthworks for lime-stabilized soil dams have been completed in other American states and elsewhere in the world. In Thailand, the Mun Bon dam was constructed with a series of lime-stabilized soil protective layers built up on the dam slope. The project used over 60,000 m³ of lime-stabilized soil (Supakij, 1995).

7.5.3 Canals

In California, following failure of a concrete lining to the Friant-Kern irrigation canal due to it being built on expansive clay, the US Bureau of Reclamation arranged for the canal to be de-watered. The sides and base of the canal were reconstructed using lime-stabilized soil and the initial intention was to resurface them with a new reinforced concrete lining. Due to the needs of agriculture, the Bureau was forced to reopen the canal before resurfacing. Over the nine months for which it was in use, the lime-stabilized soil canal was at times transporting around 100 m³/s (4,000 ft³/s) of water at a flow speed of 1.36 m/s (4.5 ft/s). Ten months later, when the canal was re-drained to continue construction, engineers found the sides and base of stabilized soil were firm and hard. Imprints of the compaction machinery and tractor tyre marks made over ten months previously were clearly visible. There was only minimal erosion to the unprotected lime-stabilized surfaces (Boynton, 1980).

7.5.4 Roads

The NLA *Lime-Treated Soil Construction Manual* (ARBA, 2004) was significantly revised for the 11th edition. It was originally written by the American Road Builders Association (ARBA) sub-committee on lime stabilization and first published in 1959 as ARBA Technical Bulletin 243 and about 90,000 copies of previous editions have been distributed. The manual describes and illustrates stages of road sub-base and base construction. Each of the steps needed to use lime-stabilized soil and pozzolans for a wide range of soil and clay types are explained. It states that:

> *Base stabilization is used for new road construction and reconstruction of worn-out roads, and generally requires adding 2 to 4 per cent lime by weight of the dry soil. In-situ road mixing is most commonly used for base stabilization, although off-site central mixing can also be employed. Lime is also used to improve the properties of soil/aggregate mixtures in full depth recycling.* (ARBA, 2004: 6)

7.5.5 Embankments

An example of dealing with difficult site conditions by using lime stabilization and pozzolan is given by Buxton Lime Industries Ltd, in their project report for building the A13 bypass near Dagenham, Essex (Buxton Lime Industries, 1990). The soil was of silt overlaying 7 m depth of soft clay. Laboratory testing using different proportions of lime (lime base) and pulverized fuel ash (PFA) confirmed that the site subsoil could be stabilized to meet the DoT Class 2 specification for highway works. Buxton reported that 100,000 m³ of soil was stabilized in layers to form the embankment on which a £71,000,000 (£71 million) section of the bypass was built (Buxton Lime Industries, 1990).

7.5.6 Housing and community areas

Setting out roadways and pavements, together with excavations for mains services and footings, can create difficult site conditions – particularly in wet weather. Site traffic can be slowed or prevented from working by deep mud. These conditions can be mitigated by an initial modification of the soil with lime followed by subsequent applications of lime for additional stabilization treatment and drying. Excavated material can be limed as it is removed and stored to mature. After mellowing, it can be returned and compacted in lifts as a road base. In large developments, parking areas for shopping centres and large stores can be included in this process.

An example of one of the many projects completed in the UK is Taylor Wimpey's Gateshead residential development. The sub-contract for soil stabilization was awarded to Thompsons of Prudhoe and TR Stabilization in October 2012. The mix design specified that a CBR of 5% was to be achieved and laboratory tests confirmed that this could be met by incorporating 2% quicklime in the soil. The area was excavated and replaced in 300 mm layers. Quicklime powder was applied to the working area using a tractor-mounted spreader. A Wirtgen mixer was used to mix the lime in with the soil layer. Compaction was completed using a Hamm Smooth Drum Roller. Compliance checks were continuously carried out as work progressed with laboratory testing to ensure the 5% CBR requirement had been met. The report notes that due to 'the fast acting drying properties of quicklime, wet-weather working at that time of the year was not an issue' (TR Stabilisation, n.d.).

CHAPTER 8
Testing and national standards

8.1 Current practice

Historic methods of using lime to stabilize soil for earth building have been largely forgotten, but its application in civil engineering developed rapidly during the second half of the 20th century and is now extensive. The result has been that national standards and specifications have not only been produced, but also revised and updated over a relatively short period. These changes have largely been in connection with road building and to keep pace with the developing science.

Several countries publish national standards for lime-stabilized soil substructure; however, there are few, if any, standards for lime-stabilized soil superstructure, particularly for earth mortars, plasters, renders, and blockwork. Despite this, there are excellent publications for building with earth only, of which there is a select list in the bibliography. Most publications on building with earth naturally focus on earth only construction, whilst those on lime-stabilized soil for civil-engineering purposes focus on substructure. The missing link between these two is the way in which lime stabilization can be used to transform diverse soil types into permanent building elements and finishes. The transformation from soil only is extraordinary in that a wide range of building elements that remain stable in wet conditions can be produced using mainly local soil, a small proportion of lime, and appropriate skills.

The variations in soil mineralogy, lime burning and pozzolan production across the world are immense, and it is therefore essential to understand local soils before they are used for construction. This understanding may come from experience, trial and error, or field or laboratory testing. Neither making assumptions about mix proportions, nor trial and error, are practical ways to produce satisfactory lime-stabilized soil building elements with an unknown soil. Methodical field testing and recording of all appropriate materials, stabilization mixes, and building components are therefore the most suitable way forward and provide a method of immediate implementation.

Chapter 2 describes areas where lime-stabilized soil has been used historically and in connection with like-for-like repairs. In the past, building with local materials was the norm, and knowledge of the way in which materials performed was gained by experience or handed down from generation to generation.

Few of those dealing with earth construction today are able to benefit from these advantages and therefore they need to rely on archive documentation, trial and error, or preferably, in respect of immediate application, methodical testing. (See photo 19.)

8.2 Field testing

Materials may be tested in the field, in the laboratory, or both.

Where there is an urgent need to rebuild, particularly during disaster relief and recovery, immediate field testing of materials, rather than relying on uncertain traditions or limited experience, may well be the best option. As well as providing rapid results, field testing can be done on site at negligible cost, using very little equipment. Another advantage of field testing is that those carrying it out soon develop an understanding and appreciation of local building materials, their best application, and skills in their use.

Chapters 4, 5 and 6 recommend carrying out field tests in three stages:

1. Testing individual materials for quality.
2. Testing mixes for stabilization.
3. Testing building elements for performance.

The field tests proposed in these chapters are practical ways in which to establish whether the quality of materials available, the properties of possible mixes, and the performance of each item tested are satisfactory. Guidelines for target levels of acceptability are taken from existing national standards for building limes, pozzolans, and lime-stabilized soil as well as other publications on building with earth listed in the bibliography.

In most cases, testing in accordance with national standards requires laboratory facilities and equipment and a controlled environment to ensure the level of accuracy expected. Field testing is therefore appropriate where this is not possible. However remote, laboratory testing in accordance with national standards should also be undertaken when accessible and could be in parallel with or following the field tests in order to verify and provide more detailed records. (See photos 17a, 17b and 19.)

8.3 Lime reactivity

Section 4.1 sets out field tests to establish lime quality. If the lime is defective or has been poorly burnt it will not be sufficiently reactive, and so more of it will be required to stabilize the soil, or, in some cases, it may be completely unable to do so.

Industrialized large-scale lime production is usually regulated to ensure that lime is produced in accordance with national standards. The regular laboratory testing required is seldom possible for small-scale lime producers in remote areas but the field-test procedures and results described in this book are based on principles similar to those widely accepted by current national standards for building limes and soil stabilization.

A select list of standards from the USA, Britain, Europe and India for building limes is given below. These standards specify and detail laboratory tests for a range of requirements, including reactivity.

American standards

ASTM C110-76a Methods of Physical Testing of Quicklime, Hydrated Lime, and Limestone
ASTM C977-18 Standard Specification for Quicklime and Hydrated Lime for Soil Stabilization

British and European standards

BS EN 459-1 Building Lime – Part 1: Definitions, Specifications and Conformity Criteria
BS EN 459-2 Building Lime – Part 2: Test Methods

Indian standards

IS:712 Specification for Building Limes

8.4 Soil suitability

Field tests for soil are principally to identify soil by particle size and determine the approximate proportions of gravel, sand, silt, and clay, and are given in Section 4.2. Related national standards for soils include:

- BS 1377:1990: Methods of Test for Soils for Civil Engineering Purposes
- *Code of Practice for Rammed Earth Structures*, IT Publications, 1996
- BS 5930:1981/2020: Classification of Soils for Engineering Purposes
- BS 5930:2015 + A1: 2020: Preparation of Soil for Testing

8.5 Aggregate selection

Section 4.3 describes simple field tests for sand, mostly to determine if the particle-size grading is appropriate for particular building elements: well-graded sand generally gives the best results. Standards for sand are often connected with mortars and renders, provide useful information and give important guidance for laboratory testing. Since recent standards are generally related to the use of sand with Portland cement-bound mixes, the list below includes some older standards which give particle-size gradings more applicable to mixes using lime binders.

Table 8.1 Aggregates

BS EN 13139:2002	Aggregates for Mortar
BS 1199 and 1200:1976	Specification for Building Sands from Natural Sources
BS 4551:1972	Sand for Mortar Testing (withdrawn and out of print)
BS 882:1992	Specification for Aggregates from Natural Sources for Concrete
BS 410	Specification for Test Sieves
BS 812	Methods of Sampling and Testing of Mineral Aggregates, Sands and Fillers
ASTM C144-18	Specification for Aggregate for Masonry Mortar
AASHTO No. M45-70	American Association of State Highway and Transportation Officials Standard
ASTM C136-04	Standard Test Method for Sieve Analysis of Fine and Coarse Aggregates

8.6 Pozzolans

Pozzolanic material is included in a mix to supplement or increase the hydraulic set of lime-stabilized soil, particularly for those soils with a low clay content. A pozzolan reacts with the lime in the mix to form insoluble compounds. Generally, the finest particle sizes of the pozzolan will be the most reactive: if the particles are too large, reactivity is reduced and may be negligible. The larger particles tend to act as additional aggregate but not as pozzolan. Field tests for pozzolanic reactivity are given in Section 4.4. Establishing the level of reactivity and quantity of pozzolan required may indicate that transporting it from further afield would be worthwhile and cost-effective. There are many manufacturing activities associated with agriculture and industry that produce waste that may be useful as a low-cost pozzolan, particularly from combustion processes: for example, blast-furnace slag and pulverized fuel ash (PFA), although less toxic materials from agricultural waste are preferable. Field research and analysis of pozzolan are best supported by laboratory testing in accordance with appropriate national standards selected from those listed below.

American standards

ASTM C311-77	Standard Methods of Sampling and Testing Fly Ash or Natural Pozzolanas
ASTM C593-06	Fly Ash and Other Pozzolans for Use with Lime for Soil Stabilization (reapproved 2011)
ASTM C821-78-14	Lime for Use with Pozzolanas (reapproved 2000).

British and European standards

BS 3892:1996 – Part 2	Pulverized Fuel Ash for Use in Concrete
BS EN 196-5	Pozzolanicity Test for Pozzolanic Cements
BS EN 14227-4	Fly Ash Specification (for Hydraulically Bound Mixtures)
BS EN 15167-1	Ground Granulated Blast-furnace Slag Specification

Indian standards

IS : 1727	Methods of Test for Pozzolanic Materials
IS : 1344	Specification for Calcined Clay Pozzolana (Burnt Clay)
IS : 3812 - 1 (part 1)	Specification for Fly Ash for Use as a Pozzolana
IS : 4098	Specification for Lime Pozzolana Mixture

Netherlands standard

N 488-11D:691.5	Definition and Specification of Trass – for Use as a Pozzolan (refer to section 4.4.2 page 86)

8.7 Stabilization

The basic principles for estimating lime-stabilized soil mix proportions and testing these on a small scale are similar to those for the construction of road sub-bases with large-scale industrialized plant. Whilst the use of lime-stabilized soil for housing and associated building elements is new (or being rediscovered), the science and technology of using lime-stabilized soil in civil engineering is well advanced. Knowledge gained from research and development, particularly that associated with road construction, has been applied on a large scale in recent decades. Examples of successfully completed engineering projects are numerous and have been carried out worldwide, some of which are described in Chapter 7. This extensive and successful application over many years inspires confidence in the use of lime-stabilized soil, and when combined with historic records (Chapter 2) offers a scientific explanation of why it has been an effective and appropriate means of building durable shelter and fortifications in the past and from at least the 13th century. The full potential of the material may be realized when traditional uses of lime and soil for general building construction are supported by advanced laboratory technology and equipment for materials analysis, testing, and monitoring. Appropriately applied and detailed lime-stabilized soil is suitable for the construction of a building's superstructure comprising a wide range of building elements, from foundations to finishes.

The national standards described here are therefore a helpful guide to the principles of lime stabilization which may be applied to projects at a variety of scales and levels of complexity – from civil engineering for international airport runways, to mortar for the smallest of homes.

British and European standards

BS 1924:1990	Stabilized Materials (Methods of Test for Stabilized Soils)
BS EN 13286-50	Proctor Compaction for Indirect Tensile and Compression Testing
BS EN 14227-15:2015	Hydraulically Bound Mixtures – Specifications Part 15: Hydraulically Stabilized Soils

American standards

ASTM D6276-19	Determination of Lime Demand (Eades-Grim Test)
ASTM D3551	Lab Preparation of Soil–Lime Mixtures Using a Mechanical Mixer
ASTM D5102	Test for Unconfined Compressive Strength of Compacted Soil–Lime Mixtures
ASTM D3877	Standard Test Methods for One-dimensional Expansion, Shrinkage, and Uplift Pressure of Soil–Lime Mixtures
ASTM D698	Test Methods for Laboratory Compaction of Soil Using Standard Effort
ASTM D1557-12e1	Test Methods for Laboratory Compaction of Soil Using Modified Effort
AASHTO T 99-19	Standard Proctor Test
NLA – National Lime Association USA	Technical Brief: 2006 Mixture Design and Testing Procedures for Lime-stabilized Soil

New Zealand specification

NZTA M15 Notes: 2012	NZ Transport Agency – Notes on Specification for Lime for Use in Soil Stabilization

Conclusion

'Seek and ye shall find.'

Not only is lime-stabilized soil one of the most ecologically sensitive building materials on the planet, but confidence in application of such an environmentally benign material is gained from observing centuries-old historic buildings that incorporate it in their construction. Many excellent examples exist across the world. Modern history too, during the last half of the last century to the present day, includes industrial development with massive civil engineering projects using this amazing material.

Lime-stabilized soil has a well proven record as an environmentally benign material that is used for road and runway construction, flood control and disaster resilience. It is suitable for self-build and low cost housing at scale in rural areas as detailed here. The challenge now for those in the building industry, is to take the development of this material to another level. There are many historic buildings that exemplify how successful it can be when employed to express permanence, culture and tradition. This includes the finest decoration and finishes. There is great opportunity to learn from the past, applying that experience with current scientific research in the use of lime-stabilized soil in an urban context. The magnificence and beauty of many historic buildings standing today are themselves evidence that this is entirely possible and a challenge to be met and embraced. Such a challenge is particularly relevant now, when the environmental impact of building materials needs to be a foremost consideration. The incidence of extreme weather and flooding globally is likely to continue to increase with climate change. This highlights the need not just for effective, durable and appropriate measures to reduce the devastating impacts on those communities in high-risk areas, but also to ensure that those measures themselves do not add to the construction industry's already carbon-intensive burden on the environment. Lime-stabilized soil offers such a measure, both mitigating and adapting to climate change, while honouring the vast and wonderful diversity of regional and cultural vernacular traditions.

The science and technology of building with earth and lime may be implemented at a range of levels between the extremes of elementary, and the highly technical, but its successful use ultimately depends on understanding and selecting the optimum method of applying available resources. We acknowledge the contribution of Schumacher to this and hope he would recognize the spirit of 'an intermediate technology' so clearly set out in *Small is Beautiful* (Schumacher, 1973).

APPENDIX 1
Establishing quicklime proportions

These lime–soil proportions will vary depending upon soil type but are a guide for evaluating initial trial mixes. All percentages give the volume of lime as a percentage of the volume of soil in the mix.

From linear shrinkage test results

Table A1.1 Proportions of powdered quicklime to soil volume for trial mixes.

Shrinkage of tempered soil (without lime) in a mould 600 mm × 40 mm × 40mm (2' × 1½" × 1½")	Estimated clay content of soil	Volume of quicklime powder to add as percentage of soil volume	Ratio of quicklime to soil	Suggested ratios of quicklime to soil for test mixes (make at least 3 specimen cubes of each mix)		
< 12 mm	1–2%	12–15%	3–6%	1:33–1:17	1:30	1:20	1:15
12–23 mm	2–4%	15–20%	6–8%	1:17–1:12	1:15	1:14	1:12
24–35 mm	4–6%	20–25%	8–10%	1:12–1:10	1:12	1:11	1:10
36–48 mm	6–8%	25–30+%	10–12%	1:10–1:8	1:10	1:9	1:8

For details of this test method refer to Section 5.3.

Table A1.2 Suggested proportions of crushed **quicklime** to soil volume for use in trial mixes.

Shrinkage of tempered soil (without lime) in a mould 600 mm × 40 mm × 40 mm (2' × 1½" × 1½")	Suggested proportions of quicklime to soil for test mixes (make at least 3 specimen cubes of each mix)		
< 12 mm (½")	1:30	1:20	1:15
12–23 mm (½"–1")	1:15	1:14	1:12
24–35 mm (1"–1⅜")	1:12	1:11	1:10
36–48 mm (1⅜"–2")	1:10	1:9	1:8

Using lime to clay fraction

Another starting point to establish the optimum proportions of quicklime for soil stabilization is to first determine the percentage of clay in the soil sample then prepare trial mixes based on adding an amount of powdered quicklime that is equivalent to 20% of the clay fraction (i.e. a fifth of the clay percentage) as follows:

- For a soil that is 15% clay, use a volume of quicklime that is 3% of the soil volume.

- For a soil that is 20% clay, use a volume of quicklime that is 4% of the soil volume.
- For a soil that is 30% clay, use a volume of quicklime that is 6% of the soil volume.
- For a soil that is 40% clay, use a volume of quicklime that is 8% of the soil volume.

This results in the use of a slightly smaller quantity of quicklime than that proposed for trials by the linear shrinkage method above.

By adjusting soil pH

Details of a laboratory test method for determining an approximate lime to soil ratio for stabilization is given in ASTM C997-10 and is known as the Eades and Grim test. It relies on adding lime to soil samples to achieve a pH of between 12.3 and 12.4 (strongly alkaline). Paper reagent strips to test the pH can be used in the field (see also Photo 13), although greater accuracy can be expected under laboratory conditions.

Further details regarding relevant national standards are given in Chapter 8.

Alternative methods

As an alternative, to confirm sufficient quicklime has been added for stabilization, carry out the field tests described in Section 5.8: testing cube, block, and disc specimens of stabilized building elements after 28 days curing by immersion in water, and for wet compressive strength. Carry out further tests, with reduced lime proportions, to establish the minimum amount of lime required for full stabilization.

For further information on test methods, refer to *Building with Earth* (Norton, 1997), *Earth Construction* (Houben and Guillaud, 1989) and *Building with Lime* (Holmes and Wingate, 2002).

See equivalent Table 5.7 for proportioning lime in the form of lime putty or dry hydrate, as alternatives to quicklime. (Note that proportions of lime putty or dry hydrate to soil will be different to those of quicklime to soil.)

Table A1.3 Suggested proportions of **lime putty or dry hydrate** to use in trial mixes.

Shrinkage of tempered soil (without lime) in a mould 600 mm × 40 mm × 40 mm (2' × 1½" × 1½")	Suggested lime putty or dry hydrate to soil for test mixes (make at least 3 specimen cubes of each mix)		
< 12 mm (½")	1:15	1:10	1:8
12–23 mm (½"–1")	1:8	1:7	1:6
24–35 mm (1"–1⅜")	1:6	1:5.5	1:5
36–48 mm (1⅜"–2")	1:5	1:4.5	1:4

APPENDIX 2
Mechanized equipment

An extensive range of machinery produced worldwide for mixing and compacting un-stabilized soil can be readily used for stabilized soil. Many of these devices, including large motorized machines designed for mass production, are described in *Earth Construction* (Houben and Guillaud, 1989).

Research will determine if appropriate equipment is available locally. If not, it may be possible to encourage its development and manufacture, so aiding entrepreneurial development. Hand-powered, animal-powered or mechanically driven equipment includes the following:

Figure A2.1 Agricultural backpack sprayer.

- An agricultural backpack sprayer. This is a very useful addition for misting (light spraying) of walls or blocks with water during the curing period. Application is easier than by hand from brushes, and gives a more uniform coverage.
- Portable jaw crusher, hammer mill, ball mill, grinder, or roller mixer for crushing and grinding quicklime and pozzolans to fine powder (see Figure 4.17).

198 BUILDING WITH LIME-STABILIZED SOIL

- A chaff cutter, fibre or straw chopper (see Figure 4.44).
- Rock crusher (large jaw crusher) for large aggregate.
- Paddle mixer or other form of roller-pan or paddle mixer for mixing all materials (see Figures A2.2 and A2.3).

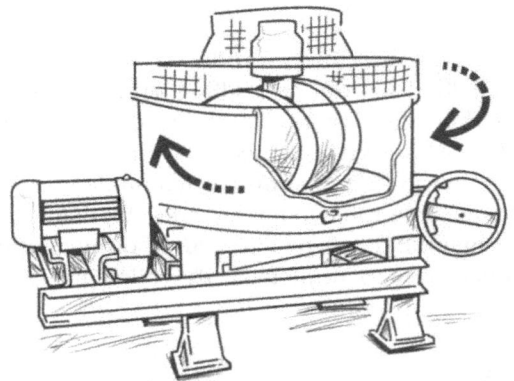

Figure A2.2 Roller-pan mixer.

Figure A2.3 Paddle mixer.

- CINVA ram, block press or equivalent as an alternative ramming method for block compaction (see Figure 6.14). The ram or block press is a steel box with a base that is raised and lowered to compact soil. This method compresses and releases the block by manual operation using a long-handled lever. It is reported that a two- or three-person team can make 300 lime-stabilized blocks in one day and a team of four or five can make 500 blocks per day per machine. Mechanically driven block-pressing machines are manufactured worldwide. These range from small output presses to motorized combined hydraulic and mechanical production plant. A comprehensive description of the many and varied types available is given in *Earth Construction* (Houben and Guillaud, 1989).

APPENDIX 3
What is lime? A geological explanation

This description of the formation of limestone is based on information given in *Building with Lime* (Holmes and Wingate, 2002).

Sedimentary rock originates from pre-existing rocks and other material. Fine-grained material eroded by water or wind is carried away from its source, largely by rivers, and deposited in lake and ocean depressions, or spread over the surface of the seabed. Limestone consists mainly, or entirely, of calcitic material produced by plants or animals that precipitate calcium compounds from water by bacterial or chemical action.

The skeletons and shells of marine animals are mainly, if not entirely, calcium carbonate. When the animals die, their shells and bones fall to the seabed and mix with accumulating inorganic sediment produced by the weathering and disintegration of material, usually from land formations. The further the inorganic material is from its origin, the greater the proportion of calcareous material in the limestone. In many cases, calcareous material can make up virtually the entire deposit, and this type of pure limestone is used in many manufacturing processes including the production of 'pure' (non-hydraulic) building lime.

Sediments that make up limestone can accumulate simultaneously with those of clay, silt, and sand. Some of these impurities are the origin of the active clay component of hydraulic limes and natural cements. One of the most favourable environments for a build-up of sediment is shallow water where there is little or no wave action. These conditions may be found near the coasts of seas and lakes, many of which have changed over geological time.

APPENDIX 4
Suitability of soils for the addition of lime

Table A4.1 Suitability of soils for the addition of lime.

Soil	Shrinkage and swelling	Sensitivity to frost action	Bulk density (kg/m³)	Voids ratio	General suitability for the addition of lime
Clean gravel, well-graded	Almost none	Almost none	2,000	0.35	Suitable for lime concrete;* the addition of sand will improve performance
Clean gravel, poorly graded	Almost none	Almost none	1,840	0.45	Suitable for lime concrete* but grading and addition of sand will improve performance
Silty gravel	Almost none	Slight to medium	1,760	0.50	Not suitable
Clayey gravel	Very slight	Slight to medium	1,920	0.40	Suitable for stabilization
Clean sand, well-graded	Almost none	Almost none	1,920	0.40	Suitable for mortars, plasters, and render
Clean sand, poorly graded	Almost none	Almost none	1,600	0.70	Suitable for mortar but grading will improve performance
Silty sand	Almost none	Slight to high	1,600	0.70	Not suitable
Clayey sand	Slight to medium	Slight to high	1,700	0.60	Suitable for daub and soil structures; suitable for use in weak render coats particularly for use with weak backgrounds
Low-plasticity clay	Medium to high	Slight to high	1,520	0.80	Suitable for stabilized road formation and stabilized earth render, but improves with the addition of sand
Organic silt	Medium to high	Medium to high	1,440	0.90	Not suitable
Clays with low plasticity	Medium to high	Medium to high	1,440	0.90	Suitable for stabilization

(continued)

Table A4.1 Continued.

Soil	Shrinkage and swelling	Sensitivity to frost action	Bulk density (kg/m³)	Voids ratio	General suitability for the addition of lime
Highly plastic clay	High	Very slight	1,440	0.90	Suitable for road stabilization and, if sand is added, for soil structures
Highly plastic silt	High	Medium to high	1,600	0.70	Not suitable
Highly plastic organic earth	High	Very high	1,600	0.70	Not suitable
Peat	Very high	Slight	1,600	0.70	Not suitable

* Limecrete is the term now preferred by the authors.
Source: UNCHS (1987) and Holmes and Wingate (2002: Appendix 8).

APPENDIX 5
Pozzolan testing

Extracts from Nederland (Dutch) Standard N 488 (1932) kindly provided and translated by Michael Wingate.

Specification for assessment

1. The material must be very dry.
2. The maximum retention on a Dutch standard (0.2 mm) sieve [...] 20%. This percentage is on the basis of the weight when the material has been dried at 100°C.
3. Minimum tensile and compressive strengths of the test pieces in kg/cm².

	Tensile strength	Compressive strength
3 (days) in the air and then 11 (days) under water	12	50 kg/cm² (5 N/mm²)
3 (days) in the air and then 25 (days) under water	16	70 kg/cm² (7 N/mm²)

The test pieces must be made up in the proportions of two parts by weight of trass, one part by weight of hydrated lime,[1] three parts by weight of standard sand with from 0.90 to 0.95 parts by weight of water.

The temperature of the air and of the water in which the test pieces are hardened must be kept as close as possible to 15°C.[2]

Comments

1. For special purposes a very finely milled trass is also available having a maximum of 10% resting on a N480-d-0.210 (0.21 mm) sieve.
2. A preliminary assessment may be made with the [Vicat] needle test described below, but this would not be the basis of an outright rejection.

To set up the needle test a stiff 'dough' is prepared by kneading together two parts by weight of trass, one part of weight of hydrated lime[1] and one part by weight of water. This dough should be thrown or shaken into a metal box, then the edges should be smoothed over and the box should then immediately be put under water at a temperature as close as possible to 15°C.[2]

After two days the mortar should be of a suitable stiffness to show no measurable impression from the needle (ground to cross-section of 1 mm²) when a load of 300 g is applied.

Notes

1. Hydrated lime is defined in N491 for the purpose of this standard but refer to Sections 4.1.7 and 4.1.8.
2. Variations from this temperature would have a marked effect on the strengths achieved.

APPENDIX 6
Example of a test record sheet

DATE PREPARED				
MATERIALS SELECTED				
COMPONENT (render, plaster, mortar, block, screed, foundations)				
LIME TYPE	QUICKLIME REACTIVITY Boil time (min)	PUTTY SPECIFIC DENSITY	HYDRATE FINENESS % passing	SOURCE & SUPPLIER
LIME QUALITY				
SOIL ANALYSIS Source	% CLAY	% SAND	%SILT	CHEMICAL ANALYSIS
				Silica %
LINEAR SHRINKAGE BOX SHRINKAGE (mm)				Alumina %
				Iron %
				LAB :
% QUICKLIME RANGE				
SAND PARTICLE SIZE ANALYSIS (sieve test)	% retained on 5 mm	% retained on 2 mm	% retained on 0.6 mm	% retained on 0.106 mm
% PASSING SIEVE				
TEST MIXES	LIME (parts)	SOIL (parts)	SAND (parts)	OTHER (parts pozzolan/fibre/dung)
TEST MIX 1				
TEST MIX 2				
TEST MIX 3				
DISCS FOR PERMEABILITY TEST AND CUBES FOR COMPRESSIVE-STRENGTH LABORATORY TESTS	NUMBER OF CUBES/DISCS	DATE PREPARED	DATE TESTED	STRENGTH (N/mm^2)
				MAX / MIN
FIELD-TEST RESULTS	PASS	FAIL	DATE	TEST LOCATION
STEP TEST				
CURING AT 28 DAYS	SHADED	WETTED	START DATE	FINISH DATE

IMMERSION TEST	DATE PLACED UNDER WATER	NO. OF DAYS STABLE
NAME OF AREA AND DISTRICT		
NAME OF VILLAGE WHERE USED		
HOUSE WHERE USED		
RECORDED BY		
PHOTOGRAPHS AND RECORDS STORED AT		
SAMPLES STORED AT		
NAME OF TEST MONITOR		
DATE	SIGNATURE	

APPENDIX 7
Compressive-strength requirements, test results and strength-gain estimates

Table A7.1 Compressive-strength requirements.

Component	Minimum compressive strength requirement at 28 days		Anticipated strength of mixes with lime binders after 2 years
	N/mm^2	lbs/in^2	N/mm^2
Internal non-load-bearing walls	1.5	220.8	3.0
Load-bearing walls for 1- and 2-storey buildings	2.75	404.8	5.5

Source: *London Building Constructional By-laws 1972* (GLC, 1973).

Table A7.2 Compressive strength of LSS samples – test results.

Sample	Compressive strength achieved		Anticipated strength after 2 years
	N/mm^2	lbs/in^2	N/mm^2
1 part moderately hydraulic lime : 3 sand at 28 days	2.6	380	5.2
1 part eminently hydraulic lime : 3 sand at 28 days	6.0	880	12.0
1 part moderately hydraulic lime : 2 crushed brick (pozzolan) at 2 years (Dibdin, 1911)	6.18	910	6.18
Un-stabilized compacted soil block at 28 days (Norton, 1997)	1.5	220.8	1.5
Lime-stabilized and compacted soil block at 14 days (confirmed following lab test 2013 – HANDS personal communication, 24/11/13)	4.08	600	8.5
Concrete penetrometer test results on numerous lime-stabilized soil trial mixes, after immersion times of 6 weeks to 6 months (IOM and HANDS personal communication, September and November 2014)	4.75	700 (penetrometer maximum reading)	6.0–10.0
UK laboratory unconfined compressive-strength test result for lime-stabilized soil from Pakistan flood-relief programme (W. Revie, Construction Materials Consultants Ltd, personal communication, April 2015)	2.4	353.28	3.6

(continued)

Table A7.2 Continued.

Sample	Compressive strength achieved		Anticipated strength after 2 years
	N/mm^2	lbs/in^2	N/mm^2
Independent laboratory test results for Arup Engineering, undertaken by NED University of Engineering and Technology, New Material Testing Facility 2017 for lime-stabilized blocks (*Flood Resilient Shelter in Pakistan* – Arup, 2017)	7	1015	8.0–10.0
Wet compressive strength field tests undertaken with the Technical Research Team at Tamera, Portugal, 2018–2019: various initial samples for road stabilization trial mixes after immersion for two weeks achieved the field test equipment's maximum reading of 7 N/mm^2	7	1015	8.0–10.0

Strength conversions based on 1 N/mm^2 = 147.2 psi

APPENDIX 8
Field tests checklist

STAGE 1: Test materials		
Preparation		Prepare record sheets and establish on-site recording procedures.
Investigate required materials available	1	Investigate local materials: soil; sand; lime; pozzolan; straw; cow dung; oil; water. If uncertain about soil suitability, field test with chemical reagent indicator strips, or arrange for laboratory chemical analysis. Alternatively, complete longer-term field tests to the end of Stage 2.
Soil	2	Conduct soil tests: clay-content test and particle-size (sieve) test. Select soil with a clay content and particle-size distribution suitable for the building element concerned.
Lime	3	Test quicklime fresh from the kiln and no more than one week after this; observation test; weight test; six-second test; reactivity test.
	4	Slake the quicklime to putty if it is not to be used immediately as crushed quicklime or hydrated lime (field-test slaking tanks or smaller containers can be made simply and in a variety of sizes).
	5	Check lime putty before use to check that it has the correct density.
Pozzolan	6	Conduct a simple reactivity test using milk of lime for reactivity. For low-clay soils or sandy soils or sand, add two or more parts of pozzolanic material to one part lime to create a trial mix and check hydraulic set with immersion and compressive-strength tests.

STAGE 2: Test mixes for stabilization		
	7	Measure linear shrinkage of soil sample to establish the clay percentage.
Prepare trial mixes	8	Calculate lime to soil proportions for three different trial mixes (see Tables 5.5 and 5.7).
	9	Select appropriate materials and proportions that are expected to produce a stabilized mix.
	10	Make three samples of each of the three trial mixes for each test.
Cure	11	Dampen all trial-mix samples at regular intervals and keep shaded from direct sun and protected from heavy rain for 28 days.
Test	12	Field test trial sample mixes after 28 days curing for: • dry strength – step test mixes intended for foundations and wall bricks and blocks;

(continued)

Continued

STAGE 2: Test mixes for stabilization		
		• stabilization – immersion test **all** trial mixes using sample cubes for foundation and wall-block mixes, and sample discs for mortars, renders, plasters, and screeds; • permeability – test renders and screeds; • wet strength – penetrometer test.
Recording	13	Use a test record sheet (see Appendix 6) to monitor and record: • mix components and ratios; • curing conditions; • test results.

Stage 3: Manufacture and test building elements and modify mixes as necessary		
Select materials and mixes	14	Build *only* with materials and lime–soil mixes tested and found to be suitable in the previous stages. Modify and test mixes for each building element or component as described in Chapter 6.
Test finished elements	15	Test components again for stabilization and for workability, adhesion, and solidity in an appropriate manner: e.g. create trial render and plaster panels, test floor screed on floor slab and limewash on surface to which it is to be applied.
Monitor quality	16	Test all production-run materials, mixes, and components on a regular basis throughout the construction process.
Train operatives	17	Train operatives to use the correct materials and proportions when preparing lime-stabilized mixes and finished building elements and to carry out tests with precision.
Motivate and demonstrate	18	Explain the benefits of lime use, including comparative cost, stability and durability, environmental sustainability, health, and heritage. Demonstrate relative stability of lime-stabilized blocks and discs under water compared with strong sun-dried mud blocks and compressed-earth blocks.
Produce documentation	19	Document the process and results using test record sheets, photographs, video, and interviews with end users. Continue to monitor and document over time, if possible, particularly after heavy rain, monsoon floods, and earthquakes.

Stage 4: Laboratory test		
Laboratory test	20	To support and corroborate field-test results, if possible, send soil for laboratory analysis. Also, replicate all mixes that have remained stable after six months' immersion and send samples of these for laboratory testing. The laboratory should carry out full chemical and particle-size analysis of the successfully lime-stabilized unmodified and modified soil sample mixes.
Archive records	21	Retain records of soil analysis and fully lime-stabilized mixes in national archives. Co-ordinate with organizations carrying out (future) geological mapping of mineralogy and soil types.

APPENDIX 9
Chemical formulae

Chemical formulae of constituents (oxides, compounds, liquids and gases) that are found in limestones and soils which are related to or can affect hydraulic set and soil stabilization.

Alumina (aluminium oxide)	Al_2O_3
Anhydrite (gypsum, calcium sulphate)	$CaSO_4$
Anorthite	$CaAl_2Si_2O_8$
Calcium carbonate or calcite (limestone)	$CaCO_3$
Calcium oxide (quicklime)	CaO
Carbon dioxide	CO_2
Dicalcium aluminate silicate	$2CaO \cdot Al_2O_3 \cdot SiO_2$
Dicalcium silicate	$2CaO \cdot SiO_2$
Halite (salt, sodium chloride)	$NaCl$
Iron oxide	Fe_2O_3
Iron sulphide	FeS_2
Kaolin (clay)	$H_4Al_2Si_2O_9$
Magnesium carbonate	$MgCO_3$
Manganese oxide	MnO
Potassium oxide	K_2O
Potassium sulphate	K_2SO_4
Silica	SiO_2
Sodium oxide	Na_2O
Sulphur	S
Sulphur trioxide	SO_3
Titanium dioxide	TiO_2
Tricalcium aluminate	$3CaO \cdot Al_2O_3$
Water	H_2O
Wollastonite	$CaSiO_3$

Note on clays: Kaolin is only one of many clay types, a number of which are mentioned in Section 7.3. The chemical formula of kaolin $H_4Al_2Si_2O_9$, however, demonstrates the high prevalence of alumina and silica in clays. The importance of these minerals is that they can react with calcium (lime) to form insoluble compounds.

APPENDIX 10
Vapour permeability: lime and lime-stabilized soil

A field test for the water permeability of lime-stabilized soil render is given at 5.8.6. The ability to bring a soil to sufficient impermeability that it is able to contain liquid water, has many benefits related to the storage and control of water flows. This benefit, as demonstrated by the field test, appears contrary to another benefit of lime-stabilized soil, in that it is also vapour permeable.

Water vapour transmission is an important element of building assembly design for most walling (and roofing) systems. The difference in vapour pressure between two sides of an above-ground building envelope assembly is what drives moisture vapour through the wall from one side to the other, and many wall assembly materials and methods require either internal vapour barriers to stop moisture created inside a building from entering the wall system (where it can cause moisture damage) and/or external one-directional breather membranes, which form a barrier to water/rain, but allow for vapour transmission to the outside of the building.

Earth buildings however, are typically of deep monolithic walls (adobe, rammed earth, cob) without the gaps that can be found in cavity walling systems which alone can account for vast amounts of moisture migration due to vapour-laden air infiltration. Additionally, the hygroscopic clay element of an earthen walling material and/or internal clay plaster offers a degree of 'humidity control' enabling uptake of excess moisture vapour in high relative humidity incidence (eg hot showers/cooking), eliminating condensation & associated toxic mould growth, and by the release of excess vapour load as the internal relative humidity level drops. The extent of this humidity regulatory property of unfired clay earth and plasters, and their 'moisture buffering' ability, relies on variables such as the vapour permeability and vapour permeance of the plaster, clay type and mineralogy, particle size distribution, thickness and density, but most clay plasters of a depth of approx 25 mm (1") can generally regulate relative interior humidity to between 40% and 70% – the ideal level at which to support human health. (Building Environment: The Moisture Buffering Capacity of Unfired Clay Masonry – Volume 82, December 2014). As explained by earth building specialist Professor Gernot Minke, "unfired clay can absorb and desorb indoor humidity faster than any other building material" (Minke. G. Earth Construction Handbook, 2006).

Straw bale, cellulose and other plant-based wall systems:
The vapour permeability of render and plaster is particularly important for those wall systems which require a high vapour transfer ability, so not to trap moisture vapour within the fabric of the wall where it could condense and impair or damage its fabric and performance, such as the natural insulation materials of straw bale and other fibre based wall systems. These walls therefore have traditionally been rendered with the highest vapour permeable render mixes, which at the same time are able to provide durability against erosion from rain in non-extreme climate areas: non-hydraulic (pure) lime and sand (plus short fibre for additional tensile strength to minimise the risk of cracks).

Lime-stabilized soil render, plaster and mortar however, offer the dual advantages of allowing for varying degrees of vapour permeability whilst also providing higher degrees of water resistance for areas of risk (flood or exposure to severe weather/base and top of walls/openings). Findings described as important, by building physicist Professor John Straube, show the addition of 10% and even 50% lime to a clay plaster mix did not significantly reduce the vapour permeability of the plaster (*Moisture Properties of Plaster and Stucco for Strawbale Buildings, Ecological Buildings* – EbNet, 2002). See Table A10.1 (and test procedure note at the end of the appendix).

Comparative test results on vapour permeability values: Table A10.1 provides an interesting compilation of comparative vapour permeability values for a range of av.38 mm (1.5") plaster samples (and light-strawclay samples) based on separate permeability tests by Straube (2000, 2002) and Minke (2001) measured in ng/Pa s m. *(Note: there is no available data on the soil composition/clay fraction content of the different mixes tested, so the results are useful as an initial guide to vapour permeability comparisons).*

- The light-strawclay wall-mix samples proved the most permeable tested, (a straw-rich mix of straw and liquid clay allowed to set hard). The low density sample showed a value of 82.7, and the higher density of 41.3, clearly showing the direct relationship between increased density and decreased permeability.
- The next most permeable mixes were predominantly the earth plasters (each in Straube's samples of approximate similar density: 1,700 kg/m^3), and the results ranged from 45.7 to 40.1 (Straube), and 23.3 (Minke), where the lower permeability showing in Minke's results could well be related to a difference in density. The addition of the lime at 10% and 50% by volume to the same Straube earth plaster mixes produced results of 41.7 and 40.4 respectively, evidencing very little reduction in permeability from the addition of the lime.
- Lime:sand plasters had the highest permeability values in Straube's tests, and vary from 56.5 – a higher permeability value than achieved by the entire range of Straube' clay plaster samples – to 16.9 (Minke).

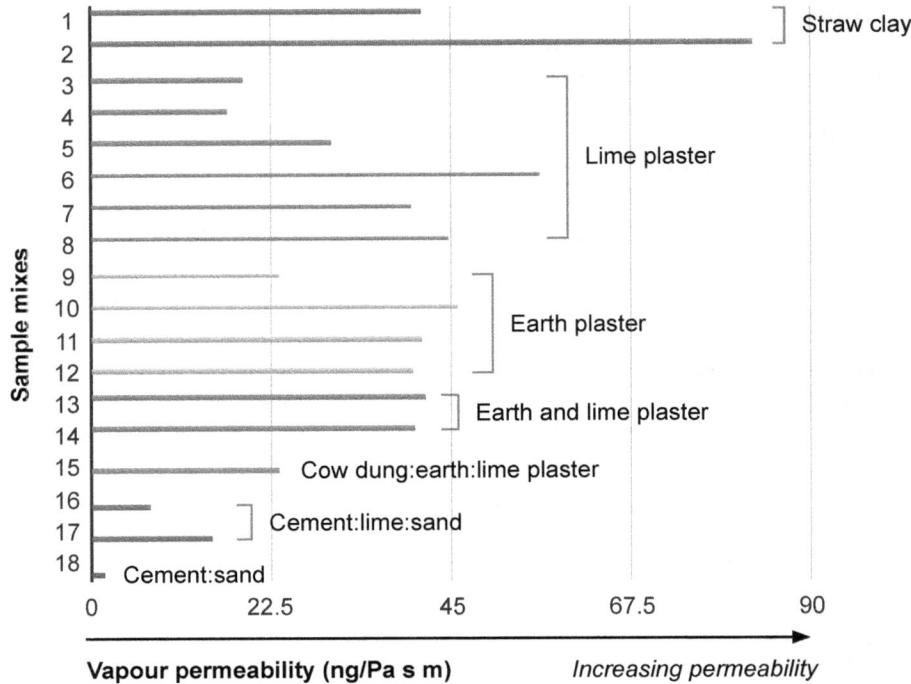

Table A10.1 Compilation of Test Results of Comparative Vapour Permeability Values of Plaster Mixes (ng/Pa s m).

1. Strawclay (1,250 kg/m³) – 41.3 (Minke)
2. Strawclay (450 kg/m³) – 82.7 (Minke)
3. Lime-sand – 18.9 (Straube 2000)
4. Lime-sand – 16.9 (Minke)
5. Lime-sand – 30 (Straube 2002)
6. Lime sand 1:3 56.5 (Straube 2000)
7. Lime:Sand 1:3 (type 'S' hydrated lime) with 5 coats of lime wash: 40.8 (Straube 2002)
8. Lime:Sand 1:3 (type 'S' hydrated lime) with wheat paste mixed with interior clay paint: 44.5 (Straube 2002)
9. Earth – 23.3 (Minke)
10. Earth 1 – 45.7 (Straube 2002)
11. Earth 2 – 41.2 (Straube 2002)
12. Earth 3 – 40.1 (Straube 2002)
13. 10% lime by volume in clay plaster mix 41.7 (Straube 2002)
14. 50% lime by volume in clay plaster mix 40.4 (Straube 2002)
15. Cow-dung earth-lime – 23.5 (Minke)
16. Cement:lime:sand 1:1:6 – 7.3 (Straube 2000)
17. Cement:lime:sand 1:2:9 – 15.0 (Straube 2000)
18. Cement:sand – 1:3 – 1.6 (Straube 2000)

Source: From Straube 2000, 2002, and Minke 2001

A (type S) lime:sand plaster finished with 5 coats of lime wash, had a finding of 40.8, which shows no notable decrease in permeability caused by the lime wash, and was found to be 'essentially as permeable as the bare earth' (Straube), a similar result to that obtained for a wheat-stabilized clay paint finish to the same plaster mix, at 44.5.
- The lowest permeability value was 1.6, for what was the cement:sand (OPC) plaster, and clearly shows that such a mix is almost vapour resistant, meeting the definition of a vapour barrier in many cases (Straube). Notably, the studies found its 'most remarkable result' to be the influence of lime in increasing permeance. The increasing addition of lime in the cement:lime:sand mixes for instance, shows 'the powerful vapour diffusion enhancement effect of lime', up from 1.6 with no lime, to 7.3 with the addition of one part of lime (in a 1:1:6 mix, Straube 2000).

Conclusions: Such test results support the suitability of various lime-stabilized soil renders and plasters as very likely appropriate finishes for such as straw bale walls in severe weather areas (prolonged wind driven incidence/ monsoon rains), where a vapour permeable render is also required to repel liquid water. In additional parallel tests conducted by Straube (2002), higher vapour permeability plasters also showed the ability to enable faster drying – another factor important for such as straw bale wall systems.

It is important to note however, that a strong hydraulic lime/high proportions of pozzolans in a render, whilst increasing the water resilience and repellency of the render, will correspondingly decrease its vapour permeability.

Testing is needed to demonstrate the comparative vapour permeability of different forms of lime, and most especially of different lime-stabilized soil render mixes, which are likely to show a direct relationship between reduction of permeability and increased hydraulic strength. Whilst also an important consideration for renders to monolithic earth walling, a decrease in vapour permeability is a particularly critical issue for such as straw bale buildings.

Vapour permeability testing for a range of lime-stabilized soil would be very welcome, and whilst such tests should be carried out under controlled laboratory conditions to international standards, scope also exists for the development of effective vapour permeability field tests.

Test procedure note

Straube and EbNet, 2002: undertaken on prepared earth and lime based plaster samples of the same thickness, made by natural building company Vital Systems with density measurements to allow for quantitative comparison. ASTM E96 formed the basis of the test, with slight modifications in test equipment to more accurately mimic the critical drying conditions for wet material, eg 100% internal relative humidity (RH), to external moderately humid conditions (75% RH). (Rather than the standard 'wet cup test' which measures 100–50%).

Sections 4.2 of the report 'Vapour Permeability, Test Equipment and Setup' describes the test equipment in detail, and 4.3 the procedure, in which samples were tightly fixed horizontally over a sealed (aluminium tape and paraffin wax) vapour impermeable container with a deep water reservoir. The area of perpendicular sample exposed to the interior and exterior of the container was the same, and each were placed in an insulated controlled temperature and humidity chamber, where humidity was controlled with a saturated salt solution in fan circulated air, and a thermostatically controlled electric heater used to maintain the chamber temperature at 25°C (77°F) within a control margin of better than +/− 1°C (+/− 1.8°F). All mass measurements were taken by an electronic Sartorius balance (capable of accurate measurements up to 12 kg with an accuracy of +/− 0.01 g). Samples were weighed before being placed in the constant temperature/humidity chamber and weighed again at regular intervals over a series of weeks. Weight change was plotted on a chart over a corresponding time interval. When the rate of weight change of each sample over at least three intervals reached equilibrium (within 2%), the vapour permeance and permeability values of the sample were calculated from the measured, steady-state weight loss (or gain).

Permeance of the sample was calculated from:
Permeance in ng/Pa s m^2 = (weight loss in nanograms)/[(duration of time interval in seconds) × (sample area in m^2) × (average vapour pressure difference in Pa)]
and **Permeability** is calculated as:
Permeability in ng/Pa s m = permeance/avg sample thickness in m.

(*Moisture Properties of Plaster and Stucco for Strawbale Buildings, Ecological Buildings* – EbNet, 2002)

APPENDIX 11
Earthquakes

(a) Lime-stabilized soil and earthquakes

Martin Hammer

Prior to using lime-stabilized soil for structural elements of a building such as foundations or walls, the project location must be evaluated for its level of seismic hazard. Global areas prone to moderate to severe earthquakes are approximately shown in Figure A11.1.

Structures in locations prone to earthquakes, especially locations subject to a peak ground acceleration (PGA) of ≥ .33 of gravity, should be designed and constructed to safely withstand design-level seismic forces without collapse. Information regarding levels of seismic hazard can be found through various sources. Governmental documents should be consulted when available. A colour global map with more detailed regional levels of expected PGA can be found at: <https://cloud-storage.globalquakemodel.org/public/Global%20Maps/version_2018.1_20200928.pdf>.

Lime-stabilized soil can potentially be used structurally in seismically active locations, but the structure should be designed by a civil engineer, or follow national building codes or established guidelines and safe practices for the region. This is particularly important for multi-storey or public structures.

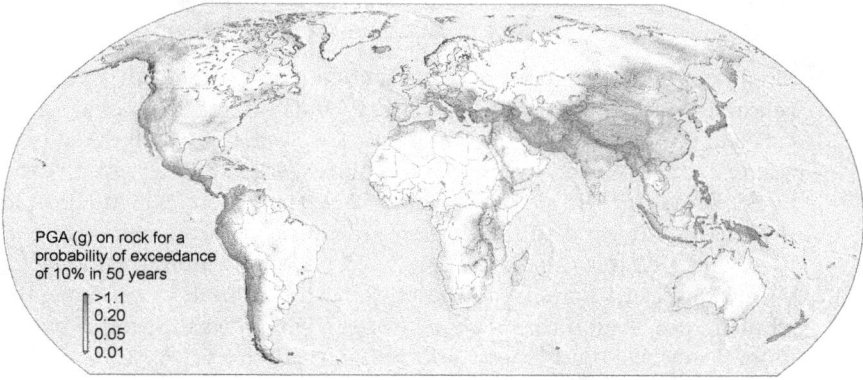

Figure A11.1 Global areas of seismic hazard
Source: Pagani et al., 2018. This work is licensed under the terms of the Creative Commons Attribution-ShareAlike 4.0 International License (CC BY-SA): https://creativecommons.org/licenses/by-sa/4.0/

216 BUILDING WITH LIME-STABILIZED SOIL

Earthquake Tips, a useful primer on earthquakes and earthquake design and construction can be found at: <www.nicee.org/EQTips.php>. Portions are India-specific but most parts are universal.

In earthquake-resistant masonry construction, it is important that the mortar be weaker than the masonry units, as explained below. Lime and lime-stabilized soil mortars offer optimal strength and ductility for use with stronger masonry units such as fired brick and stone, as compared with cement mortars that are too strong and brittle, while providing much greater durability than mud mortar. Lime-stabilized soil mortars are also appropriate with lime-stabilized masonry units when the mortar is of a weaker mix.

World Housing Encyclopedia <www.world-housing.net> offers a resource on construction in earthquake regions, including an overview of earthquake-resistant vernacular housing which can be found at: <www.world-housing.net/major-construction-types/vernacular-introduction>.

Martin Hammer is Architect and Co-director of Builders Without Borders www.builderswithoutborders.org

(b) Earthquakes and lime and lime-stabilized mortars for masonry construction

Prof. Randolph Langenbach, FAAR

The following is a series of excerpts from the book *Don't Tear it Down, Preserving the Earthquake Resistant Vernacular Architecture of Kashmir* and several research papers by Randolph Langenbach, which focus on different types of traditional masonry construction where the strength of the mortar is discussed. The book can be purchased at <www.traditional-is-modern.net/> and the papers cited here are available open access online at <www.conservationtech.com> (see Bibliography and references below).

> Much of this research was inspired by an Indo-American Exchange Fellowship to Srinagar, Kashmir, India in 1981, with the discovery of ... 4, 5 and 6 story masonry and timber houses, some of which dated back centuries, and which survived an earthquake in 1885 as reported on by British visitor Arthur Neve, who wrote: 'The general construction in the city of Srinagar is suitable for an earthquake country; wood is freely used, and well jointed; clay is employed instead of mortar, and gives a somewhat elastic bonding to the bricks ... If well built in this style the whole house, even if three or four stories high, sways together, whereas more heavy rigid buildings would split and fall' [Neve, A., 1913. Thirty Years in Kashmir, London]. (Langenbach, 1987, 2009)
>
> Earthquakes over the most recent decades in Pakistan, India, Turkey, Italy and other places have shown remarkably similar results which have

proved the advantage of [relatively] weak (i.e., mud & lime) rather than strong (i.e., cement) mortar in masonry construction.

An important ... factor in the performance of the walls [is that the] bricks be stronger than the mortar. The mud or weak lime mortar tended to encourage sliding along the masonry bedding planes instead of cracking through the masonry units when the masonry panels deformed ... This pattern of damage helped explain why these buildings were capable of surviving a major earthquake that had felled modern reinforced-concrete buildings [as happened by the thousands in the 1999 Turkey earthquake] ... thus: strength and rigidity are less effective than flexibility, ductile behavior, and cumulative nondestructive damping. (Langenbach, 2003)

The use of weak mortar serves to protect the masonry units from fracture so that the ultimate wall strength is the masonry crushing strength ... One reason why engineers have failed to recognize the benefit of modifying the infill masonry in RC buildings, or to include contributions from it in their calculations, is the reliance on linear elastic models in seismic design, therefore ignoring in an overly conservative manner any post-elastic strength and energy dissipation. (Langenbach, 2008)

The *hımış* and *dhajji-dewari* structures that have been found standing amidst the earthquake ruins of the modern [reinforced concrete] buildings around them do not mock their modern neighbors laid low, but rather, they quietly encourage us to shed ... the arrogance and over-confidence that brought the newer structures into being, forcing us to re-examine the roots of our civilization for ideas of how to build better in the present, even while we explore new and more modern materials and forms for the future. (Langenbach, 2006)

Randolph Lagenbach, appointed Assistant Professor of Architecture, University of California at Berkeley; Senior Analyst at the Federal Emergency Management Agency specializing in the repair and upgrading of buildings after earthquakes; consultant for UNESCO and UN-HABITAT.

Quotes from R. Lagenbach papers available open source at: <www.conservationtech.com>: (1987) 'Masonry as a ductile material: traditional and contemporary construction practices utilizing unreinforced masonry in seismic areas', (2003) 'Survivors amongst the rubble: traditional timber-laced masonry buildings that survived the great 1999 earthquakes in Turkey and the 2001 earthquake in India, while modern buildings fell', (2006) '"Opus craticium" to earthquake resistant traditional construction', (2008) 'Learning from the past to protect the future: armature crosswalls'.

Glossary

Reproduced with permission for duplication, additions and amendments from the publisher and authors of *Building with Lime* (Holmes and Wingate, 2002).

Absorption. The process of taking in or sucking up a liquid or gas into a solid but permeable material, e.g. rainwater into mortar or render.

Active clay. Any clay which will produce an active pozzolan by firing at a suitable temperature or, in some cases, without firing. These are likely to be soft clays such as fine kaolinites, illites, and montmorillonites, but the mineralogy is significant.

Adobe. Spanish word for 'earth'. A method of unfired earth construction in which the clay soil is first made into blocks which are allowed to dry out before they are used in the masonry.

Aggregate. The hard filler material, such as sand and stones, in mortars, plasters, and renders.

Agricultural lime. Any lime used for soil conditioning, mainly to raise a soil's pH, but usually a term to describe ground-up chalk or limestone which is calcium carbonate. In building terms this is not called lime, although crushed or powdered limestone of an appropriate grading can make a good aggregate.

Air limes. Limes which set through carbonation rather than through chemical reaction with water.

Air-slaked lime. The mixture of calcium (and possibly magnesium) carbonate, hydroxide and oxide which results when a quicklime slakes naturally in moist air. For most purposes, this is a degraded material.

Alumina. One of the earths; the only oxide of aluminium (Al_2O_3); the chief constituent of all clays. When combined with water, it forms a colloidal hydrate, $Al(OH)_3$. An important pozzolanic component in active clay.

Aluminates. In this context, compound of alumina with one of the stronger bases (e.g. calcium aluminate). See Alumina.

Artificial pozzolan. A human-made material which will react with lime and water to give a hydraulic set. For example: reactive brick dust, or ash from fuels used in some industrial processes.

Autogenous healing. The self-healing of fine cracks in a mortar or render due to the binder already in that mortar. Moisture transports free lime into the cracks.

Background. The masonry, lathing, or other surface on which the plaster or render coats are built up.

Backing coat. The first of two or more coats in a plaster system.

Bagged lime. Usually quicklime lumps or dry hydrate of lime. Calcium (and perhaps magnesium) oxide or hydroxide sold in sacks. Bagged lime may refer to bagged quicklime in lump or granular form.

Batch. Prescribed quantities of constituents making up a single load for mixing.

Binder. The material which forms the matrix between aggregate particles in a mortar, plaster, render, or lime concrete. It is a paste when first prepared but must then harden to hold the aggregate in a coherent state. Examples include lime and clay.

Blinding. A covering of soft material, usually sand, to level out the (hardcore) uneven surface below, to provide support for a damp-proof membrane (DPM) and/or screed.

Bond.
1. The overlapping of stones, bricks, or other masonry units in a wall or other structure.
2. The adhesion between two surfaces: for example, a render and its backing.

Breathability. The extent to which a building material is able to allow moisture to move to the surface and evaporate harmlessly.

Building element. An essential part of a building: for example, foundation, blockwork, or render.

Bulk density. The mass per unit volume of a liquid or solid including the voids in a bulk sample of the material, expressed in national standards by g/ml.

Bund. Wall or enclosed area usually formed to encase or protect against spilt or excess liquid.

Calcareous (material). Material containing any form of calcium carbonate or lime.

Calcination. The process of isolating calcium oxide or pure calcium from the parent material. In this context, usually by burning limestone.

Calcite. The mineral form of calcium carbonate which gives strength to a well-carbonated lime mortar. It occurs naturally as rhombohedral crystals of Iceland spar and is able to produce double refraction of light; this is the reason for the exceptionally bright appearance of limewashed surfaces.

Calcium aluminate. A compound formed by the combination of calcium hydroxide ($Ca(OH)_2$) and aluminium oxide. In this context, as one of the reactions after mixing lime hydrate with active clay (or other aluminates) in soil and in the presence of water. The reaction relates to forming a cementitious binding material during the process of stabilizing soil with lime. This can result in a hydraulic set over a period of time depending on local conditions.

Calcium carbonate. $CaCO_3$, the material from which lime is prepared. Natural forms are limestones, chalks, shells, and corals. It is also formed as an industrial by-product in, for example, acetylene manufacture. Mortars, renders, and plasters containing calcium hydroxide take up carbon dioxide from the air to form calcium carbonate, and this develops the set.

Calcium hydroxide. $Ca(OH)_2$, slaked lime or hydrated lime, also lime putty and milk of lime.

Calcium oxide. CaO, commonly called quicklime.

Calcium silicate. A compound formed by the combination of calcium hydroxide ($Ca(OH)_2$) with silica. In this context, as one of the reactions after mixing lime hydrate with active clay (or other silicates) in soil and in the presence of water. The reaction relates to forming a cementitious binding material during the process of stabilizing soil with lime. This can result in a hydraulic set over a period of time depending on local conditions.

Carbonation. Carbonation is the process of forming carbonates and, in this context, particularly the formation of calcium carbonate from calcium hydroxide when a lime develops its set. A lime mortar is said to have carbonated when the binder has reacted with carbon dioxide from the air and developed strength beyond that which is achieved simply by drying out.

Casein. A protein in milk with many industrial applications including glue making. It can form an adhesive with lime.

Cathode. A negatively charged electrode.

Cation. A positively charged ion that is attracted to a cathode during electrolysis.

Caustic (substance). A substance that burns or corrodes organic matter or tissue.

CBR test. A penetration test to evaluate the subgrade strength of flexible roads and pavements by assessing the California bearing ratio of undisturbed, remoulded and compacted soil specimens. A standard Proctor mould is standard equipment for carrying out this test.

Cement.

1. **Portland cement:** A quick-setting binder for making mortars and concretes. The chemical set of cement takes up most of the free lime, losing many of the benefits gained from using building limes for materials where very high strengths are not required. The traditional description of the most common and widespread cement is ordinary Portland

cement (OPC) or just Portland cement. It is formed by fine grinding a clinker which has been prepared at a very high temperature from a mixture of clay and limestone with small amounts of gypsum added to control early setting. Modern Portland cement is not recommended for use here due to the disadvantages outlined above as well as the amount of CO_2 and toxic wastes produced by its manufacture not reabsorbed during set. When used for mortars, plasters and renders, its inflexibility and impermeability can be harmful to the adjacent building fabric.

2. **Natural cement:** Other forms of cement include natural cements which historically were prepared by burning calcareous clay or calcareous clay nodules or rock. Natural cements were used extensively in the 19th century and are currently produced in various forms in Europe and the USA. They usually provide good weathering qualities and have good workability but are expected to be of less strength than OPC. They were superseded by Portland cement during the 20th century, largely due to the demand for greater strength.

3. **Roman cement:** The first quick-setting natural cement produced from nodules known as septaria, found initially in English river estuaries. Patented by James Parker in 1796, the material was widely used for over 50 years. It has a characteristic warm brown colour and usually received a paint finish. In addition to render it was extensively used for decorative moulding, including run and cast work and hand modelling. It is no longer commercially produced in England.

4. **Cementitious:** A description that applies to a wide range of materials including Portland cement, natural cement, Roman cement, hydraulic lime and non-hydraulic lime. The term 'cement' was used historically in connection with many materials including building limes, before the advent of Roman and Portland cements.

5. **Hydraulic lime:** A distinction between cement and hydraulic limes is that cements are burned at higher temperatures and must be ground to a fine powder before they can slake. Refer to **Hydraulic limes**.

Chalk. A common, white, soft and porous calcium carbonate rock from the cretaceous age with a very fine structure.

Chunam. A fine plaster based on very pure or shell lime and fine aggregate which is highly polished and widely used in India for the highest quality finishes, often to external walls and roofs. Also referred to as *arayash* and *sudha*.

Clay. The smallest particles produced by the weathering of rocks: each particle is less than two microns across. Chemically, clay particles are mostly hydrated aluminosilicates and, physically, they are usually in the form of thin plates which stack together.

Clinker. A hard solid material formed by the fusion of materials at high temperatures. It is detrimental to all lime and most lime-based mixes.

Coarse stuff. A mixture of lime putty and aggregate, usually sand, which is stored to mature for use as a plaster, render, or mortar.

Cob. One method of construction for earth walls. Soil plasticized with water is mixed with straw and placed on the wall top with a fork or by hand. It is compacted into place and any excess is pared off with a cutting tool.

Coherent (state). The condition of a mixture when the different materials of which it is formed have stuck together and remain united.

Compressive strength. In this context, the property of a hardened mortar or lime-stabilized soil to resist compressive forces usually expressed in newtons per millimetre squared (N/mm^2), the equivalent megapascals (Mpa), or pounds per square inch (lbs/in^2).

Concrete. A structural building material which can be cast in a fluid state or rammed in a moist state, but will set to a firm solid. It consists of sands and stones with water and in this context possibly with soil and/or pozzolan and a lime binder. (A concrete differs from mortar and render in that it usually has a higher compressive strength and larger aggregate sizes.) Also see **Cement**.

Conservation. In this context, action to secure the preservation or survival of buildings or building material.

Coral ragstone. Ragstone extracted from the ground (limestone) formed over geological time by coral (not coral taken from the sea).

Core (in quicklime). Under-burnt calcium carbonate (limestone) within a lump of quicklime which has not fully converted to calcium oxide due to insufficient time or temperature in the kiln.

Cure (to cure). The setting and hardening of a plastic mix containing a binder. See **Curing**.

Curing. In this context, the process of keeping a lime-stabilized soil mix under specific environmental conditions (providing appropriate levels of moisture, temperature and protection) over sufficient time for carbonation and chemical set to achieve the desired properties for intended use.

Damping down. Lightly misting or wetting.

Dead-burnt lime. Calcium oxide formed at extremely high kiln temperatures. It has a dense physical structure which does not allow it to hydrate under normal conditions, although it may hydrate eventually causing defects in finished work.

Deionise. Action that causes a reduction of ions.

Density vessel. A vessel of set (standard) dimensions to contain a specific volume of material, usually in liquid or powder form, so that a measurement of mass allows the density of the material to be established.

Distemper. A historical paint or paint medium where the binder is of vegetable- or animal-based glue. An abrasion-resistant 'hard distemper' may include materials such as soap, casein or linseed oil.

Distil. To vaporize and re-condense liquid.

DPM. Damp-proof membrane. (Sometimes referred also as DPC: damp-proof course.)

Drag (hair hook or hoe). A tool like a broad-pronged rake or hoe with a long handle. In this context, used for raking quicklime and putty to turn it over at the bottom of a slaking tank.

Durability. The ability of a construction or material to resist degradation by environmental conditions and decay mechanisms throughout its design life.

Earth bedding/bedding mortar. Mortar for bedding units, used when differentiating from pointing mortar.

Electrode. A conductor through which electrical forces enter or leave an object, substance, or region. In this context, it is used in connection with electrostatic forces bonding micelles of clay during lime stabilization.

Eminently hydraulic lime. Lime prepared from a limestone containing a proportion of active clay. A suitable binder for work that must withstand extreme conditions including flooding. Unlike a natural cement, eminently hydraulic lime contains enough free lime to enable it to break up and slake to powder when water is added. Natural cement, on the other hand, must be finely ground before it can hydrate.

Fat lime. Lime putty, sometimes used in connection with other forms of lime when hydrated, usually non-hydraulic, having a good workability.

Feebly hydraulic lime. Slightly hydraulic lime containing a small proportion of active clay, typically less than 12%. It should set in water in 15 to 20 days, although it may take longer.

Float (to float).
1. A float is a laying-on and smoothing tool for plastering (see Figure 4.2, Item 17).
2. The action of laying plaster or other materials onto an existing surface and smoothing it.

Flocculation. The gathering together or clotting of fine particles in a dispersed state to form large agglomerations; here used in connection with pozzolanic reactivity.

Fly ash. A very fine coal ash which may have pozzolanic properties. See also **PFA**.

Fossil. The recognizable remains or impression of a usually prehistoric plant or animal preserved in a stratum of earth (in this context, limestone or calcareous material).

Fraction. Numerical quantity that is not a whole number. In this context, the proportion of a mixture or soil composed of a particular substance or mineral.

Free lime. Lime in a mortar which remains as calcium hydroxide and has not yet carbonated or combined with a pozzolan. It may be transported by moisture within the pore structure and can be available for deposition to heal fine cracks, known as autogenous healing.

Grading. The proportion by mass of an aggregate in various size fractions.

Grout. A fluid mortar used to fill joints that are too small to access with mortar of normal consistency.

Gypsum ($CaSo_4 \cdot 2H_2O$). The dihydrate of calcium sulphate from which the various forms of gypsum plaster are prepared by dehydration, and to which they revert on setting.

Hair hook (drag or hoe). A tool like a broad-pronged rake for mixing hair into a lime plaster or mortar.

Hand-picking. The process of handling large lumps of quicklime to assess their densities and rejecting the denser lumps as these are either over-burnt or contain an overly large core of unburnt material.

Harling. A thrown finish of lime and aggregate similar to roughcast, common in Scotland.

Hot mix. In this context, lime-stabilized soil mixes prepared by slaking quicklime in soil and combining them in the hot state while the lime slakes. The amount of water and dwell time before use is varied due to the wide range of soil types. This requires initial (field) testing and an understanding of local conditions. Hot mixes can be with quicklime as small lump lime, granulated quicklime or quicklime powder. The quality and form of quicklime selected is determined by requirements of the building element for which the mix is prepared.

Hoe. See **Drag**.

Hydrated hydraulic lime. Lime which has been hydrated into a dry powder. It will set under water and is sold to users in bagged form.

Hydrated lime; dry hydrate. $Ca(OH)_2$, calcium hydroxide. When it is produced from quicklime, the heat of hydration drives off any excess water to leave a dry white (or lightly coloured) powder. The powder may be improved by removing heavy particles using a sieve or air separation. It may be stored in a silo for some days or weeks to allow for late hydration before being packed in containers (usually paper or plastic sacks) or distributed in tankers if it has been mass-produced in a mechanized plant.

Hydrated lime may be of non-hydraulic lime $Ca(OH)_2$ or hydraulic lime, but the heat of hydration and speed of slaking for hydraulic lime is less than that for non-hydraulic lime.

Hydration. In this context the absorption of water and its chemical combination with calcium oxide (quicklime) to form calcium hydroxide (lime putty or dry hydrate).

Hydraulic binder. A binder which sets and develops strength by chemical interaction with water. It can set under water.

Hydraulic limes. Natural hydraulic limes are prepared from limestones or chalks with clayey impurities. Artificial hydraulic limes are manufactured by mixing calcium hydroxide with a pozzolan which enables the limes to harden, even in damp conditions. Hydraulic limes were originally used in areas subject to frequent saturation or continuously damp conditions, and for hydraulic engineering works such as harbour walls, dams, bridge piers, and canal embankments.

Hydraulic set (of limes). The chemical combination of lime, unburnt clay, burnt clay or other pozzolanic material and water to form a stable compound, even under water. This can either be arranged by mixing an air lime (or any other) with a clay soil or pozzolan and water, or by mixing a hydraulic lime and water. In the latter case, the lime and clay will already have formed intermediate compounds in its formation or in the firing.

Impervious. Forming a barrier to water in its liquid state.

Ion. An atom or molecule with a net electric charge due to the loss or gain of one or more electrons.

Ion crowding. An increase in the number of ions.

Jaghery, jaggery. A coarse dark brown Indian sugar made from the sap of the jaggery palm tree. It is diluted with water and added in small quantities to the mix for a mortar, or render, or particularly to strengthen the finishing coat in a plaster. Sugar greatly increases the solubility of lime in water.

Kankar. A pozzolanic ingredient of limes and aggregates used for hydraulic mixes in India.

Keratin. A protein found in horn, wool, hoof and hair. Traditionally used for the production of glues and sealants but sensitive to moisture.

Key. In this context, a mechanical bond produced by the physical interpenetration of the first plaster coat with the background, or of one plaster coat with another. On a lath background, applying pressure with the float when laying on the first coat forms nibs between and behind the laths that become the key. The nib is ideally a dovetail shape but, in practice, can be an irregular nib.

Kiln. A furnace or oven in which material is heated to effect chemical reaction. In this context, the calcination of lime by driving off carbon dioxide from calcium carbonate to form calcium oxide. A lime kiln may be constructed of various materials including lime-stabilized soil. Significant factors in kiln design and choice of materials are heat resistance, insulation and draught control.

Knocking-up. Mixing. In this context, combining materials thoroughly to produce a mix for building elements such as mortars, plasters and blocks.

Laminar structure. A structure composed of a series of laminates.

Laminate. Lamina. Thin plate, scale, or flake.

Large lump lime. Quicklime supplied in pieces that are over 200 mm in size and are thus suitable for hand-picking.

Larry. A long-handled tool for mixing lime putty and coarse stuff. It is like a hoe with a half-moon-shaped hole in its blade.

Late hydration. Normally, reactive quicklime combines with water to a large extent within a matter of minutes and entirely within a few hours. But quicklime burnt at a higher temperature, perhaps at a hot spot within the kiln, and some of the more hydraulic limes may be less reactive. It may not combine with water until some days or even months later. This can be a serious problem since the lime in its hydrated state may be almost double the volume of the quicklime. This situation can give rise to defects in the finished work (often referred to as 'popping' or 'pitting').

Laterite. Soil formed as the result of tropical weathering of rocks rich in iron, alumina, or both; often with varying proportions of free silica, quartz, and clay. Typically, the iron fraction, which can vary between 55% and 40%, is one of the principal causes of hardening when exposed to air and/or dried. This hardening quality is used in tropical countries to produce sun-dried bricks and earth structures that harden without firing or the use of other stabilization methods, although they can dissolve in wet conditions without stabilization. The addition of lime and/or compaction often has the effect of stabilizing and improving the durability of this soil as a construction material.

Lath (lathing). The riven or sawn wood which is nailed up to form the background in some plaster or daub systems. The word lathing is also used to describe alternative materials which do a similar task.

Lean lime. A lime prepared from a limestone or chalk containing impurities which do not contribute to a hydraulic set, but act instead as part of the aggregate in a mortar.

Leather hard. An indication of the density of a material, in this context soil, for testing consistency. Judged by the soil's similarity to the density of soft leather.

Le Chatelier test. A simple test to determine the soundness of lime hydrate. A sample is placed in an expandable cylindrical mould that has two 150 mm long indicators, and this is then placed in a steam cabinet to reduce the reaction time of the lime hydrate sample. The distance between the indicators is measured and determines the degree of expansion, and hence level of soundness, which may be compared with that allowed under the appropriate standards (Holmes and Wingate, 2002: 231–232).

Lime. The root meaning of the word is a sticky substance, and from that it has developed several later meanings. In this context, the word lime includes all the oxides and hydroxides of calcium (and magnesium) but excludes the carbonates. Thus quicklime and slaked lime we would call lime, but chalk and limestone we would not call lime. (Farmers, however, still use the term 'lime' when spreading crushed chalk on their fields.)

Lime concrete. See **Limecrete**.

Limecrete (sometimes referred to as lime concrete). A building material cast from a mixture of aggregate (usually sand, gravel and/or broken stone), in a matrix of hydraulic lime or of lime and pozzolan, not using Portland cement.

Lime cycle. A concept to warm the heart of environmentalists. When lime is used in buildings, it eventually reverts to calcium carbonate – the chemical from which it was originally prepared. The majority of carbon dioxide driven from the limestone during lime burning is thereby eventually replaced with carbon dioxide absorbed from the atmosphere. The full cycle is the conversion of calcium carbonate to calcium oxide (giving off carbon dioxide), the combination of this with water to form calcium hydroxide, and, finally, the carbonation of calcium hydroxide in which water is lost and carbon dioxide gained to re-form calcium carbonate.

Lime pit. A covered tank or pit used to store lime putty in moist conditions.

Lime putty. Slaked lime stored in an excess of water. This process makes the resulting putty more workable and also enables less reactive particles to be hydrated before use. In Roman times, lime putty intended for plaster of the highest standard had to be stored for three years before use. A distinction should be drawn between putty prepared from dry hydrate and putty run directly from quicklime. There is overwhelming evidence that the latter is greatly superior for applications where excellent plasticity and good carbonation are needed, as those who have mastered the craft of lime plastering will testify.

Limestone. Any rock or stone whose main constituents are calcium carbonate or calcium and magnesium carbonates.

Limewash. A simple form of paint prepared from lime, with or without various additives. It may be applied both internally and externally, and is most suitable for use on walls and ceilings.

Lump lime. Calcium oxide (quicklime) in lump form, rather than crushed to powder.

Marl.
1. A soft calcareous rock mostly made up of shells and usually containing appreciable quantities of clay and sand.
2. A subsoil of similar mineralogy to the rock but not necessarily containing shells.

Masonry. Assemblage of units of brick, block or stone set in mortar.

Matrix. An embedding or enclosing mass or mixture.

Mellow. Softened by age. In this context, part of the tempering process.

Micelles. In this context, fine-grained clay particles made up of silica and alumina molecules in thin sheets. The size of a clay micelle is in the order of 0.01 to 1 micron across.

Milk of lime. A free-flowing suspension of hydrated lime in water in such proportions as to resemble milk in appearance.

Mineralogy. The study of natural inorganic substances (minerals).

Moderately hydraulic lime. Lime which might be expected to set under water in about six to eight days, but it may take longer to do so.

Mortar. Any material in a plastic state which can be trowelled and becomes hard in place. It can be used for bedding and jointing masonry units.

Mortar mill. This operates in a similar way to a roller-pan mill. There are many variations including models which use stone rollers and troughs of varying sizes in lieu of a pan. It may be animal-, machine-, or human-powered.

Natural cement. A fortuitous balance of chalk and clay, as in certain mudstones, can have the ideal proportions to produce a quick-setting cement when fired at its optimum temperature. Like other cements, natural cements need to be finely ground before they can slake with water (hydrate). Also refer to Cement.

Natural hydraulic lime. Hydraulic lime produced by burning a mixture of naturally occurring argillaceous limestone. In BS EN 459-1 they are given the designation of NHL together with a numerical value depending on their hydraulic reactivity (see NHL 2, 3.5 and 5). Also refer to **Hydraulic limes** and **Cement**.

NHL 2, NHL 3.5, NHL 5. The European standard classification for hydraulic limes related to the minimum compressive-strength requirements of mortars expressed in N/mm^2 at 28 days.

Non-hydraulic lime. A lime of high purity, air lime, which is also loosely described as pure lime or fat lime.

Nib. In this context the plaster key, usually on lath background, made by an irregular or dovetail shape formed between and behind the laths by pressure with the float (trowel) when laying on the first coat.

OPC. Refer to **Cement**.

Over-burnt quicklime. If lime is burnt at too high a temperature the lumps contract, become less reactive and will eventually sinter, particularly if there is sufficient clay in the stone. This gives them a wizened appearance. Over-burnt lumps will not slake readily and late hydration may lead to problems in materials made using them.

Pargeting or parging.
1. Rich decoration on external plasterwork by modelling the surface.
2. Lining a surface or flue with a mix of lime putty and cow dung or other mortar mixes.

Particle-size distribution. An assessment of the proportions of material of different sizes within a sand or aggregate. For good work with lime, an even distribution is helpful.

Penetrometer (pocket penetrometer). A small hand-operated mechanical device for measuring the compressive strength of materials and components.

Permaculture. A set of design principles, agricultural systems and methods that integrate human activity with natural surroundings to create highly efficient community resilience and self-sustaining ecosystems.

Permeability. The ease with which a liquid or vapour can pass through a solid material.

PFA. Pulverized fuel ash is a waste product from power stations burning pulverized coal. Its characteristics vary depending on the coal and burning conditions, but some PFAs are pozzolanic. All are contaminated with sulphates, some much more so than others, and this may present a health hazard. The reactive parts will combine to set with lime (or with free lime in a cement), in a hydraulic reaction which can take place in locations where carbonation would not otherwise happen easily. PFA is used with lime in grouts.

Phenolphthalein indicator. A widely available chemical indicator which can clearly show the difference between neutral and alkali conditions.

It remains colourless on calcium carbonate whilst on calcium hydroxide it changes to deep purple-red. It can thus be used to show the extent of carbonation in lime mortar and lime-stabilized soil.

pH scale. A scale that expresses the acidity or alkalinity of a substance as a value between 0 (extremely acidic) and 14 (extremely alkaline).

Pigments. Colouring material which, when added to liquid, may create a paint. Most are supplied in the form of a very fine dry powder.

Pisé-de-terre. An earth-walling construction technique using compaction (or 'ramming') of a relatively dry and lean clay-content mix between robust shuttering.

Pitting and popping. A delayed defect in plasterwork caused by late hydration of over-burnt quicklime incorporated into a plaster. As the quicklime hydrates, it tries to expand and pressure builds up behind the surface. This may be released with a small explosion as a cone-shaped section of plaster in front of the speck blows forward (pops) leaving a small crater (pit) in the surface.

Plaster. Plaster is any material used in a plastic state to form a durable finishing coat to the surfaces of walls and ceilings and other elements of a building. Typical materials, in this context, are based on soil only or lime with sand, or soil, or pozzolan, or any combination of those.

Plasticity. In this context, the property of a soil or modified soil that enables it to be easily shaped and applied to a surface in its original or modified state.

Plinth. A widened above-ground section at the base of a wall.

Plumb bob. A pointed bob or weight, usually of metal, which is carried on a string or plumb line and used as an instrument for determining perpendiculars. It may be used in conjunction with a square and level to establish the horizontal.

Pointing. The finished surface layer in the joints between masonry units.

Portland cement. Refer to **Cement**.

Pozzolan, pozzolanic material. A pozzolan is any material containing constituents (generally alumina and reactive silica) which will combine with hydrated lime at normal temperatures in the presence of moisture to form stable insoluble compounds with binding properties. It may be used to give a hydraulic set to a mortar made with lime, or to combine with the free lime in a naturally hydraulic lime to increase its durability. There are many naturally occurring pozzolanic materials, such as certain volcanic ashes, and several artificial materials, such as crushed soft-burnt bricks. These contain clay particles composed principally of alumina, silica, and sometimes iron, which have been rendered active by heat. Fine grinding increases their reactivity.

Pozzolana. A pozzolanic material of volcanic origin from the Pozzuoli district near Mount Vesuvius, Italy.

Pozzolanicity. The extent to which a material can combine with calcium hydroxide and water at normal temperatures to produce compounds which can set and develop strength under water.

Precipitation. The deposition of moisture from the state of vapour following cooling, especially in the formation of rain. A precipitate is that which is so deposited and may result in the formation of solid material. Also the production of an insoluble solid as the result of a chemical reaction between liquids. In this context, the deposition of free lime (sometimes associated with its subsequent carbonation and/or crystallization).

Protection. In this context, protection during the curing period. Covering over the finished work to resist the effects of ambient conditions over-wetting, freezing or drying out (by sun or wind) the lime work until carbonation and/or hydraulic chemical set is sufficiently mature to be self-protecting.

Proctor mould. Equipment for testing compacted soil in accordance with the AASHTO standard Proctor test (AASHTO T 99). The mould is laboratory test equipment produced in two alternative sizes. These are hollow steel cylinders with an inside diameter of 100 mm (4″) or 150 mm (6″), used for compacting soils prior to mechanically testing compressive strength in accordance with the standard.

Quicklime. Calcium oxide, CaO. Lime which has not been slaked. Also known as lump lime, burnt lime. Called 'quick' because of its lively affinity for water. Commonly recognized sizes from ASTM C51-71 are:

- *large lump* – lumps of up to 200 mm (8″)
- *pebble* – pieces smaller than 65 mm (2½″)
- *ground, screened or granular* – particles 6.5 mm (¼″) and smaller
- *pulverized* – powder, most of which will pass an 0.850 mm (ASTM No. 20) sieve.

Raking. Dragging or scraping materials along a surface to gather them together, spread them evenly or break them up. In this context, to break up quicklime lumps and to stop them forming an unslaked mass at the bottom of a lime-slaking tank.

Rammed earth. A technique for constructing walls, foundations and floors using natural raw materials such as earth, sand, chalk or gravel, in which the constituent mix is rammed hard (compacted) within formwork. The mix is normally quite dry compared with other earthen building techniques, such as cob and adobe. See **Pisé-de-terre**.

Reactivity (of lime). The ability of lime to combine quickly in a chemical reaction. This can be seen in slaking: a reactive lime will slake immediately. It is partly dependent on the parent limestone, and largely dependent on the temperature and duration of calcination. Reactive limes have porous structures that result in a high surface area. Hard-burnt limes are less reactive. Over-burnt and dead-burnt limes are very unreactive. Pozzolans also vary in their reactivity depending on their mineralogy and the way they have been prepared.

Reagent. A chemical substance that reacts with another and is used to detect the presence of the other.

Refraction.
1. The action of breaking up.
2. In this context, the deflection of a ray of light on passing obliquely from one medium into another. Calcite crystals display double refraction: that is, light entering a crystal will be split into two beams.

Render. A formulated, external plaster system, applied in either two or three coats.

Rhombohedral. Having the form of a rhomboid: a parallelogram with adjacent sides unequal and its angles not right angles.

Roller-pan mill. An edge grinder adapted for mixing mortars. There are several variations but, in all, the mortar rests in a sturdy ring-shaped pan and is squeezed by heavy iron rollers. Either the pan rotates or the two rollers are guided around the pan. The mixing is very thorough but the grinding action may crush and change the particle-size distribution of the aggregates, depending on the way the rollers are set.

Roman cement. Refer to **Cement**.

Roughcast. A thrown dash coat of coarse aggregate mixed with wet lime binder and/or mortar to a semi-fluid consistency. The composition varies but is typically around two parts lime to three parts sand to one or two parts washed gravel or crushed stone. It may be applied to either multi-layer render or well-keyed walling.

Run-of-kiln quicklime. Quicklime just as it is drawn from the kiln, with all impurities and defects. These may include over-burnt and under-burnt material, ash, and even unburnt fuel. The quality of run-of-kiln quicklime depends on the skill of the lime burner.

Salt.
1. Common salt – sodium chloride (NaCl) extensively prepared for use as a condiment, food preservative, and for industrial processes. Soluble in water.
2. A compound formed by the union of an acid radical with a basic radical; in the binary theory of salts, any body which forms a salt with a metal or its equivalent.

3. Sodium chloride and calcium chloride salts increase the solubility of lime and delay wetting and drying cycles which benefit carbonation. Salts may be present in some soils, and can be carried by water from various sources. Sulphate-bearing salts may also be present. Expansive salts can be destructive but small concentrations may not be harmful to lime-stabilized soil construction and their effect can be mitigated to varying degrees by the addition of pozzolan.

Sand. Weathered particles of rock, usually high in silica, typically between about 0.06 mm and 5 mm (smaller than gravels and larger than silts). The particles are hard and will not crumble. Sand is used as an aggregate in mortars, plasters, and renders, as well as to modify clay soils. The properties of sand used in a mix have a major effect on its workability, final strength, and durability. The types of sand normally used in building are:

- *Sharp sand.* Consists of predominantly sharp angular grains. Clean, well-graded sharp sand is the best selection for mortar, render, and plaster as it imparts strength and durability to the finished work. Workability is improved by using fat (non-hydraulic or 'pure') lime as the binder and allowing this to stand as coarse stuff (not possible with OPC as a binder on its own).
- *Coarse sand.* A sand which is composed of predominantly large and medium-sized grains. The higher the proportion of large grains, the coarser the sand. Coarse sand is used to improve the durability of external renders and mortars. Very coarse sands usually require a lime binder, blending with other sands, or the addition of a plasticizer to assist workability. Sharp coarse sand is the most durable but the least workable, although it is suitable for roughcast.
- *Soft sand.* A sand which is composed of predominantly small and rounded grains. It often has a silt content, the proportion of which is variable. It feels soft in the hand when squeezed. The small rounded particles improve workability but can give rise to cracks and failures in the finished work.
- *Well-graded sand.* A sand with an approximately even particle-size distribution. As the smaller particles can fit in between the larger particles, this even distribution reduces the proportion of voids to solids, and thus is less demanding on the binder than poorly graded sand.
- *Blended sand.* A blend of sands of different grain sizes and sharpness to achieve a good particle-size distribution. This provides a balance between durability and workability. It is most often used in connection with plaster for backing coats and pointing mortar when the quality of available sand needs to be improved. Sand may be blended by sieving sand from a single source to adjust the particle-size proportions, or by mixing sands from different sources.

Scarify. To break up, scratch, or loosen a surface with a sharp instrument or scarifier.

Scouring. Giving a material, usually plaster, render or screed, a smooth hard surface by working it, mostly in a circular motion, with a float.

Screed.
1. In floor laying: The whole mortar layer levelled to be the finish, or to receive the finished floor surface.
2. In plastering or flooring: A carefully levelled band of mortar or plaster acting as a guide for the rule, the tool which sets the level of the whole surface.

Seismic. Relating to or caused by an earthquake, earth movement or vibration.

Self-healing. Autogenous repair of fine cracks through the precipitation of free lime.

Setting period. Time taken to solidify or harden.

Shuttering. In this context, temporary timbering or metal sheeting used for the in-situ casting of formless mixes including modified earth, lime-stabilized soil, and lime concrete; formwork.

Silica. An important mineral substance; the dioxide of silicon ($SiO2$), which in the form of quartz enters into the composition of many rocks and is contained in sponges and many plants.

Silicates. The salts of silicic acid, H_2SiO_3; the most common type of minerals. Conveniently treated as if they were compounds of silica (SiO_2) and of a metal oxide. Normally available and readily recognized in sand. An important pozzolanic component in active clay.

Silt. Fine particles 0.002–0.063 mm in a soil or aggregate. It is recommended that a soil's silt content does not exceed 20% for modified and stabilized soil mixes, and 6% for lime–sand mixes.

Sinter. Coalesce at high temperatures to form a single mass without completely melting.

Size (glue size). A glue derived from animal products and used in solution with water to reduce the suction on a porous surface before applying a paint.

Skim coat. A thin finishing coat of lime and very fine aggregates frequently used as a finishing coat to plaster or render.

Slaked lime. Calcium hydroxide, $Ca(OH)_2$. Prepared by hydrating quicklime in an excess of water to form milk of lime or lime putty, or by hydrating quicklime with minimum water to form dry hydrate of lime powder.

Slaking.
1. Slaking to putty is the action of combining quicklime with excess water to form milk of lime or lime putty.

2. Slaking to dry hydrate is the action of combining quicklime with the minimum amount of water to form dry hydrate powder.
3. Air slaking is the exposure of quicklime to sufficient air to promote hydration. This method of slaking will reduce the binding qualities of the lime as carbon dioxide is absorbed at the same time.

Soil (earth). Material predominantly formed by the weathering of rocks. It contains a mixture of clays, silts, sands, stones, and other materials. Topsoil usually contains organic matter.

Stabilization. In this context, the lime stabilization of soil. The addition of lime to clay soils produces three principal reactions:

1. When in water, calcium substitutes for alkali elements such as sodium and potassium in the soil. The clay particles flocculate to form coarser agglomerates of clay which can be compacted more readily to give an increased compressive strength.
2. Over time, calcium combines chemically with the silica and alumina in the clay mineral. In the presence of moisture, complex aluminium and calcium silicates are formed in a low-grade pozzolanic reaction that is accelerated at higher temperatures. The product is an insoluble binding material comprising insoluble calcium silicate and silica gel.
3. Alongside the two reactions described above, carbonation occurs. The lime reacts with carbon dioxide from the air to form carbonated cements.

A different but related stabilization process occurs with hydraulic limes. Natural and artificial hydraulic limes, where active clay and lime are combined before burning, are mainly compounds of calcium, silica, and alumina (principally dicalcium silicate, tricalcium silicate, and tricalcium aluminate). These are insoluble and materials containing them will, with appropriate preparation and curing, remain set under water.

Straw bale building. A method of building with dry straw stems baled into compressed blocks or prefabricated panels.

Subgrade. Usually the compacted native soil below the sub-base of a road or pavement which requires minimum compressive-strength values depending on use. The precise definitions of subgrade and sub-base vary from one country to another but in this context refer to lime-stabilized soil layers usually in connection with road construction.

Suction. In this context, the drawing in of pore water from a mortar, render or plaster into a substrate. This can assist the bond but, if not controlled, can result in rapid loss of workability of the mix or even failure.

Sulphate. A salt of sulphuric acid: usually with a term indicating the base (e.g. calcium sulphate $CaSO_4$, gypsum).

Surkhi. A softly burnt clay which is ground together with lime in a mortar mill. This acts both as a porous aggregate and a pozzolan and makes particularly good mortar.

Taghery. A shallow circular bowl or dish, similar to a wok and generally made from pressed sheet metal. It is mostly used for culinary purposes but, in this context, it may be used for small-scale field testing of building materials.

Tallow. A clarified animal fat prepared from the hard fat from the kidneys of ruminants (animals that chew grass and vegetation as their staple diet) by melting it, separating the fat from the residue.

Tamper. An instrument, usually hand operated, for compacting materials and mixes in moulds for testing.

Tempering. In this context:

1. To bring a material (soil or soil mix) to a proper state and consistency by mixing, possibly with additional water, and working up before use.
2. To allow the same material to stand undisturbed (in a wet state) to assist infusion and development of the soil or mix before use.
3. A process combining 1 and 2 above.

Tending. Caring for mixes and finishes, usually during the curing period, to ensure they are well finished, protected and dampened.

Tensile strength. Resistance to breaking under tension, or when stretched by forces pulling against each other.

Titration. Adding to a solution of known proportions a suitable reagent of known strength, until a point is reached at which reaction occurs or ceases in order to ascertain the amount of a given component in a mixture or compound.

Trass (and tuff). A pozzolanic material of volcanic origin from Germany, as a local name for tuff, a consolidated fine-grained volcanic ash from the valleys of the Eiffel, ground to a fine powder and extensively exported for use as a pozzolanic additive to mortars in the 18th and 19th centuries.

Unconfined. Not limited or restricted. In this context, a mortar sample for testing which has been demoulded and is not confined by the mould in which it was cast when tested.

Under-burnt quicklime. Quicklime which has not received enough heat to convert the whole lump from carbonate to oxide, leaving a core of unconverted material at the centre.

Unsoundness. In some mechanical processing flows, over-burnt and under-burnt material may be reintroduced into the dry hydrate or lime mixes. If such material is used for plastering without sufficient maturing as a putty (or if a putty run straight from quicklime is used too soon), then the over-burnt particles will hydrate within the body of the plaster or other mixes causing a general expansion. This breaks the bond between the plaster and the backing and can lead to hollows behind the plaster and a breakdown of finished surfaces. The Le Chatelier test shows if this will be a problem.

Vapour permeability. See Appendix 10.

Voids. In this context, the space left between sand grains and/or aggregate particles which has to be filled with a paste of binder and water to achieve a mix that is workable and durable.

Volcanic tuff/tufa. Volcanic ash used in this context as a natural pozzolan. The finest grained is called dust tuff; grains up to 2 mm ($^{1}/_{16}''$) diameter sand tuff; and 2 mm to 64 mm (2" and ½") grains are known as lapilli.

Water-burnt lime. If the slaking is badly handled, temperatures may rise too high and a hard, gritty form of calcium hydroxide may be produced.

Wattle and daub. Lightweight interlaced laths, battens or sticks woven between upright staves firmly fixed in place (wattle) plastered on both sides, usually with an earth-based render (daub).

Weathering qualities or properties. The durability of a material against the destructive actions of the weather and atmosphere. These actions include cycles of heating and cooling, frosts, wind abrasion and airborne chemicals.

Well graded. See **Sand**.

Wet-sieving method. The use of water to wash particles through a sieve. Mostly used when the sieve aperture size is too small for dry sieving without difficulty.

Workability. The ease with which a mortar plaster or similar mixes may be used. This important property is not easily defined, but it includes high plasticity and good water retention. Much sharper aggregates, which contribute to long-term durability, can be used with a highly plastic binder, such as lime putty, which provides good workability, rather than with a cement binder (OPC), which is not as plastic and more difficult to work.

Bibliography

3D design-development-display OEG (1992) *Research of the Old Dispensary – Zanzibar*, Aga Khan Cultural Services (Zanzibar) Limited, Zanzibar.

ARBA (2004) *Lime-treated Soil Construction Manual, Lime Stabilization and Lime Modification Bulletin*, 326, National Lime Association, Arlington, USA.

Arup (2017) *Flood Resilient Shelter in Pakistan – Evidence Based Research*, Ove Arup and Partners International Ltd, London.

Ashurst, J. and Ashurst, B. (1988) *Practical Building Conservation. Vol. 2: Terracotta, Brick and Earth*, English Heritage Technical Handbook, Gower Technical Press, Aldershot.

Beetham, P. et al. (2015) *Lime Stabilization for Earthworks: A UK Perspective*, revised paper (EngD) submitted at Loughborough University, UK.

Bevan, R. and Woolley, T. (2008) *Hemp Lime Construction*, IHS BRE Press, Watford.

Boynton, R.S. (1980) *Chemistry and Technology of Lime and Limestone*, John Wiley and Sons, New York.

Building Research Establishment (1980) *Overseas Building Notes*, Nos 154 and 184, HMSO, Watford.

Buxton Lime Industries (1990) *Soil Stabilization Specifiers Guide*, Triad, Manchester.

Copsey, N. (2016) *Earth Mortars and Earth Building*, Transcriptions and correspondence, 9 January 2019, Malton, Yorkshire.

Cowper, A.D. (1998 [1927]) *Lime and Lime Mortars*, BRS, HMSO, London, Reprinted Donhead Publishing.

Department of Transport Highways Agency (2001) *Specification for Highway Works Concerning Lime Stabilisation of Subgrades in the UK*, HMSO, London.

Dibdin, W.J. (1911) *The Composition and Strength of Mortars*, RIBA, London.

FAA (2011) Standards for Specifying Lime Treated Subgrade (30 Sept 2011) AC 150/5370-10F Item P-155.

Fathy, H. (1986) *Natural Energy and Vernacular Architecture*, University of Chicago Press, Chicago and London.

Fawcett, J. (1998) *Historic Floors: Their History and Conservation*, Butterworth Heinemann in association with ICOMOS UK, Oxford.

Fernandes, M. (2008) *Earth Mortars and Earth–Lime Renders*, Heritage Conservation Issue 8, University of Coimbra, Portugal.

GLC (1973) *London Building Constructional By-laws 1972*, Greater London Council, London.

Gutschick, K.A. (1985) 'Canal lining stabilization proves successful', *Pit and Quarry*, pp. 58–60.

Gwilt, J. (1894) *An Encyclopaedia of Architecture*, Revised by Wyatt Papworth, J.B. Longmans Green and Co., London and New York.

Hill, J. (1920) *Sukkur Barrage Project 1919. Vol. 1: Report on Proposed Barrage on the Indus at Sukkur, Sind*, Times Press, Bombay.

Hinnells, M. (2010) *A Low Carbon Economy: New Business Models in the Built Environment*, Environmental Change Institute, Oxford.

Historic England (2015) *Practical Building Conservation: Earth, Brick and Terracotta*, Ashgate Publishing, Farnham.

Holmes, S. (2002) 'Motslow Hill Cottage, Warwickshire', *Journal of the Building Limes Forum*, 9: 61–64, Edinburgh.

Holmes. S. and Rowan, B. (2015 [2014]) *Lime Stabilized Soil Construction: A Manual and Practical Guide*, Strawbuild, Edition 1: HANDS, Karachi; Edition 2: IOM, Islamabad and Geneva.

Holmes, S. and Wingate, M. (1989–1991) *Reports to ITDG and STCDA for Emergency Repairs, Traditional Lime Technology and Small Scale Lime Production, Zanzibar*, Rodney Melville and Partners, Leamington Spa.

Holmes, S. and Wingate, M. (2002) *Building with Lime*, Practical Action Publishing, Rugby.

Houben, H. and Guillaud, H. (1989) *Earth Construction*, Practical Action Publishing, Rugby.

Innocent, C.F. (1916) *The Development of English Building Construction*, Cambridge University Press, London.

IPCC (2019) *Global Warming of 1.5 °C. An IPCC Special Report on the impacts of global warming of 1.5 °C above pre-industrial levels and related global greenhouse gas emission pathways, in the context of strengthening the global response to the threat of climate change, sustainable development, and efforts to eradicate poverty*, Intergovernmental Panel on Climate Change (IPCC), Geneva.

IRC (2008–2013) *B/CS Unified Specifications, Section 321129, Lime Stabilization*, Industrial Resources Council (IRC), USA.

Keable, J. (2011) *Rammed Earth Structures: A Code of Practice*, second edition, Practical Action Publishing, Rugby.

Kholucy, S. (2013) 'Chimney parging', *Journal of Building Limes Forum*, 20: 66–71, Edinburgh.

Langenbach, R. (1987) 'Masonry as a ductile material: traditional and contemporary construction practices utilizing unreinforced masonry in seismic areas', *Proceedings: Fourth North American Masonry Conference*, vol. 1, Masonry Society, Los Angeles, pp. 33.1–33.14.

Langenbach, R. (2003) 'Survivors amongst the rubble: traditional timber-laced masonry buildings that survived the great 1999 earthquakes in Turkey and the 2001 earthquake in India, while modern buildings fell', *Proceedings of the International Congress on Construction History*, Madrid, Spain.

Langenbach, R. (2006) '"Opus craticium" to the "Chicago frame, earthquake resistant traditional construction', plenary address paper, in P.B. Lorenço,

P. Roca, C. Modena, and S. Agrawal (eds), *Proceedings, Structural Analysis of Historical Constructions Conference*, New Delhi.

Langenbach, R. (2008) 'Learning from the past to protect the future: armature crosswalls', *Engineering Structures*, 30(8): 2096–100.

Langenbach, R. (2009) *Don't Tear It Down! Preserving the Earthquake Resistant Vernacular Architecture of Kashmir*, UNESCO, New Delhi.

Little, D.N. (2000) *Evaluation of Structural Properties of Lime-stabilized soils and Aggregates*, The National Lime Association, Arlington, USA.

Madu, R.N. (1977) 'Investigation into the geotechnical and engineering properties of some laterites of eastern Nigeria', *Engineering Geology*, 11(2): 101–25.

McCann, J. (1983) *Clay and Cob Buildings*, Shire Publications, Aylesbury.

Mendonça de Oliveira, M. (1990) *The Study of Accelerated Carbonation of Lime-Stabilized Soils*, Sixth International Conference on Conservation of Earthen Architecture, Les Cruces, New Mexico.

Mileto, C. and Vegas, F. (2014) *La Restauracion de la Tapia en la Peninsular Iberica, (1998–2010)*, Valercia Augmentum, Portugal.

Minke, G. (2000, 2006) *Earth Construction Handbook: The Building Material Earth in Modern Architecture*, Wit Press.

Nash, G.T. (2000) *Terra Britannica*, James and James (Science Publishers), London, p. 4.

Neve, R. (1969) *The City and Country Purchaser and Builder's Dictionary: or the Complete Builder's Guide*, David and Charles (Publishers), Newton Abbot (facsimile of the 1726 edition).

Norton, J. (1997) *Building with Earth*, Practical Action Publishing, Rugby.

Ola, S.A. (1977) 'Limestone deposits and small-scale production of lime in Nigeria', *Engineering Geology* (Netherlands), Volume 11, Number 2 (June), Amsterdam.

Pagani, M., Garcia-Pelaez, J., Gee, R., Johnson, K., Poggi, V., Styron, R., Weatherill, G., Simionato, M., Vigano, D., Danciu, L. and Monelli, D. (2018) *Global Earthquake Model (GEM) Seismic Hazard Map (version 2018.1 - December 2018)*, <https://doi.org/10.13117/GEM-global-seismic-hazard-map-2018.1>.

Pearson, G.T. (1992) *Conservation of Clay and Chalk Buildings*, Donhead Publishing, London.

Reid, K. (1989) *Panel Infillings to Timber-framed Buildings*, Technical Pamphlet 11, Society for the Protection of Ancient Buildings, London and Margate.

Salmerón, P. (2007) *The Alhambra: Structure and Landscape*, new edition, La Biblioteca de la Alhambra, Granada.

Schumacher, E.F. (1973) *Small is Beautiful*, Blond and Briggs, London.

Sunshine, P. (2008) *Wattle and Daub*, Shire Publications, Oxford.

Straube, J (2002) *Moisture Properties of Plaster and Stucco for Strawbale Buildings*, Ecological Buildings, EbNet.

Supakij, N. (1995) *Infrastructure Development in Civil Engineering*, Paper No. GT 10, Kasetsart University, Bangkok.

TR Stabilisation (n.d.) *Case Study: Gateshead Project* [online] <http://www.trstabilisation.co.uk/cmsfiles/PDF/Gateshead%20Case%20Study.pdf> [accessed 25 November 2019].

UNCHS (Habitat) (1987) *The Basic Parameters of Soil as a Construction Material*, Earth Construction Technology, Technical Notes, No. 11, Part 1, United Nations Centre for Human Settlements (UNCHS), Nairobi.

Vicat, L-J. (1837) *A Practical and Scientific Treatise on Calcareous Mortars and Cements, Artificial and Natural*, J.T. Smith (translator) John Weale, London, Reprinted Donhead, Shaftesbury 1997.

Walker, B., McGregor, C. and Little, R. (1996) *Earth Structures and Construction in Scotland*, Historic Scotland, Edinburgh.

Weismann, A. and Bryce, K. (2006) *Building with Cob*, Green Books, Totnes.

William, E. (2010) *The Welsh Cottage*, Royal Commission on the Ancient and Historical Monuments of Wales, Aberystwyth.

Williams-Ellis, C. Eastwick-Field, J. and Eastwick-Field, E. (1947) *Building in Cob, Pisé and Stabilized Earth*, revised and enlarged edition, Country Life Limited, London.

Wingate, M. (1985) *Small-scale Lime-burning*, Intermediate Technology Publications, London.

Wolfe Murray, M. (2008) Embodied Energy and Carbon in Construction Materials [online] <http://www.circularecology.com/nuqdjaidjajklasah.html>.

Woolley, T., Kimmins, S., Harrison, P. and Harrison, R. (2001) *Green Building Handbook*, E. and F.S. Spon Press, London.

Index

Page numbers in *italics* refer to figures and tables.
Photos in the colour section are indexed in square brackets.

additional materials 87
advantages *see* benefits and advantages
Africa
   East 4, 177
   Fort Jesus, Kenya, 22, [photo 1b]
   Zanzibar, 13, 17, 19, [photos 1a, 5a, 5b, 16b]
aggregate 3, 30, 32
   building components 34, 105, 107, 123, 141, 143, 151–2, *151*, 172, 178
   crushing 38, *52*, 198
   for floors and screeds 23, 162–5
   for foundations 124, 126–7
   hot mixes 46
   selection *83*, 95, 159, 189–90, *190*
   stabilization 83, 98, 113, *114*
   standards 189, *190*
   sub-soil 72–3, *81*, *see also* sand
Ahmed, S.T. xvii
airport runways 177–8, *184*, 191
alkalinity 6, 45, *see also* lime-stabilized soil: pH; soil: pH test (ASTM) 103–4
alumina 27, *28*, 32, 56, 83, 218
American Road Builders Association (ARBA) 90–1, 180, 185
American Society for Testing and Materials (ASTM) 11
   aggregate selection *190*
   alkalinity test 103–4
   block composition 16
   bulk density levels for hydrated lime classification *61*
   Eades and Grim test 93, 192
   lime reactivity 189
   pozzolans 190
   soil suitability 189
   stabilization 192

Arup Consulting Engineers London 1, 9, 119, *206*
ash (as pozzolan) xv, 2, 3, 32, *83*
   see also pozzolans
Ashurst, J. and Ashurst, B. 17

Bailey, N. and Worlidge, J. 5, 18
barium chloride 92
   see also reagent strip tests
batching 57, 59, 107, 140, 171, 219
beeswax 23–4, 38, 120, 159, 165
Beetham, P. et al. 93
benefits and advantages
   ecological 5–8, 193
   economic and environmental 1–2, 7, 14
binders 7, 29, 169, 189, 219
binding properties 15, 27, 51, 54, 70, 72, 76–7, 87–90, 112
blinding 164, 219
block making area 37
block moulds 107–8
   removing block from 136, *137*
blocks
   average composition (ASTM) 16
   care in handling 137
   checking mix 136, *137*
   curing 138
   compaction 115, 135
   machine made 115, 135, 140, [photo 14]
   preparing for final placement 38–9, 41, *105*, 107–8, 134–5, 139
   stacking 138, *139*
   trial mix 104–5, *107*
Boynton, R.S. 184–5
Brazil (Bahia) 16, 18
Breese, J. xvii

brick dust 3, 32, 83, 113–14
   see also pozzolans
bricks and blocks 9–10, 132–40
   foundations 128–30
   history 17–18
   pit linings 65
   see also block moulds; blocks
British and European standards (BS EN) 189–91
British Lime Association (BLA): technical guidance 183–4
British Standard (BS)
   aggregate selection *190*
   road building 177, 180, 183
   soil suitability 189
building elements and components testing 9, 123, 219
bulk density 60–1, 219
bund 129, 219
Burford, M. xvii
Buxton Lime Industries Ltd, England 10, 185

calcareous material 199, 219, 221, 224, 228
calcination 219, 226, 232
calcite 28, 68, 159, 209, 219, 232
calcium
   aluminate 14, 27–8, 219–20
   carbonate 25, 50, 89, 115, 220
   chloride 90
   hydroxide 25, 27, 50, 68, 220
   oxide 25, 50, 57, 68, 220
   silicate 14, 27–8, 220
   sulphate (gypsum) 90–1
canals, 12, 26, 179–80, 184–5
carbon dioxide 2–3, 6, 47
   field testing gain-in-weight 57
   reabsorption 7, 25, 50, 115
   see also carbonation
carbon emissions 2–3, 5–6, *7*
carbonation 9, 25, 28, 50, 115–16, 154, 168–9, 220
   see also curing
casein (milk protein) 22, 89, 172, 220
   see also distemper
cations 27–8
CBR test 179, 184, 186, 220

ceiling
   finishes 20–1, 22, 149
   plaster 14, 20–1
cement 220–1
   impermeability 215–16
   unsustainability 2, 5–9
cementitious materials 27–9, 221
chalk 18, 159, 218, 220–1
   finish 22
chemical formulae of constituents 209
choice of materials 3, 23, 32
CINVA ram 41, 108, 115, 135, 140, 198, [photo 14]
civil engineering
   early 20th-century applications 177–8
   examples of completed projects 184–6
   industrialized processes, plant and equipment 180–2
   research and development 178–80
   technical guidance 182–4
clay 6, 9, 72, 209, 221
   and loam 5, 17–18, 21, 23
   natural indicators 73
clay content
   estimation 100, *101–2*, 195
   external render test mixes 161–2
   fraction 103, 195–6, 214, 224
   field tests 74–80
      ball drop test 76–7
      cigar test 77
      disc test 78
      granularity test 75
      linear shrinkage box test 40, 79–80, 100, *101–2*
      ribbon test 78
      sedimentation test 79
      shine test 75
      smell test 76
      wash test 75
   see also clay soils
clay lump 17
clay soils 71–2, 96, 200–1
   active clays 27–9, 30, 60–2, 68, 100, 159, 199, 218
   external render 161
   foundation stabilization 126–8

historic buildings 14, 16–17, 23
  and organic materials 87–8
  plaster 147
  road building 183
  stabilization 110–13
clay swell 72, 93
  swell test 92
clinker 7, *54*, 55, 221
colour *see* pigments; reagent strip tests
compaction 115
  blocks 135
  floor finishes 162–3
  foundation mix 125–6
  industrialized process 182
  plaster 150
compressive strength xv, 9–10, 12, 119–20, 222
  and immersion test 86, [photos 15a, 15b]
  laboratory testing 33, 36
  pozzolanic strength tests 84, *86*, 159, 202
  requirements and estimates 207–8
  roads 179–84
  tests and strength gain 50, 83, 205–6
  interval testing 123
  unconfined 33, 179–80, *205*, 236
  (ASTM) UCS test of compacted soil lime mixes 192
  wet strength 1, 12, 83, 119, 196
concrete 3, 184, 222
  see also lime concrete; lime-stabilized soil concrete; limecrete
conductivity test 92
conservation (buildings, material) xiv, xvii, 9, 13, 18, 88, 222
Construction Materials Consultants Ltd 5
Copsey, N. 4–5
coral ragstone 17, 19, 159, 222
core 53–4, 222
cow dung 88–9, 148, 161–2
Cowper, A.D. 159
crystals 27, 68, 90–1
cube moulds 41, *106–7*, 108
cubes (trial mix) 105–6, *107*
  mortar mix testing 141

curing 115–16, 222
  blocks 138
  external render 158
  floors, screeds and pit linings 166
  foundations and plinths 131
  industrialized process 182
  limewash 172–3
  see also damping down; protection

damp-proof membrane (DPM) 164, 223
damping down 115–16, 128, 222
  and brushing 153, 168
  and shading 131, 132, 138–9, *140*, 142
  external render 153–5, 158
  limewash 169–70, 172–3
  mortar 142
  screeds and pit linings 166
  see also curing
dams and flood mitigation 184
Darejo, H. xvii
decoration x, 167
demoulding 136
density vessel 60, 222
Department of International Development (DFID), UK xiii, xv, xvii, 1
Department of Transport, UK 10, 178, 180
  technical guidance – roads 182
detrimental conditions 3, 6
  durability of lime-stabilized soils 179
  quicklime 55
  stabilization 90–3,
disaster recovery xiii, xiv, xvi, 1, 3, 9–11, 188
disc moulds 105, *107*, 108
discs (trial mix) 105–6, *107*
  mortar testing 141
  render and plaster 151–2
  water permeability test 120–1
distemper 22, 223
diversity
  building elements 9
  vernacular 193
durability 223
  additional 160

compaction *135*, 145, 150, *163*
curing 158
field testing 10, 33
   immersion test 116
history and long-term 3–5, 13–14, 17, 20
limewash 172
   sustainability 2, 5–9
mortar and surface finishes 22, 65, 140, 149–50
roads 179–80

Eades and Grim 179
   test 93, 192
earth bedding *see* mortars, bedding
earth bricks and blocks *see* bricks and blocks
earth floors *see* floors
earth/soil mortars *see* mortars
earth/soil plaster *see* plaster
earth walling systems *see* walling techniques (earth)
earthquake iii, 1, 4, 9, 208, 215–17
   see also seismic
ecological/environmental benefits *see* benefits and advantages
economic benefits *see* benefits and advantages
embankments 185
England
   historic buildings 4–5, 14, [photos 2, 3a, 4a, 4b, 6, 19]
      cow dung 88–9
      earth floors 16–17
      earth mortars 19–20
      pargeting 167
      surface finishes 22
      wattle and daub 18–19, 145, *146*
   see also UK
equipment
   mechanized 41, 181–2, 197–8, [photo 14]
   and tools 37–41
ettringite 90
European standards
   British and European standards (BS EN) 189–91
   Committee for Standardization (CEN) 11
Evans, J. 17

excavation of foundation trench and footings 124–6, 129, 185
external render *see* render
eyes and skin 44–5
   see also safety precautions

fats and oils 89
   see also oil; tallow
Fawcett, J. 16–17
fibres 19, 86–7, *111*, 165, 174
   external render 159, 114, 146, 159
   machinery *87*
   optimum proportions 149
   plaster mixes 123, 143, 148–9, 151, 157
   trial proportions *151*
   types and abbreviations *121*
field testing 9–10
   building elements 123–76
   checklist 207–8
      three-stage field-testing flow chart *35*
   immersion test 14, 37, 84, 92, 104, 119, 122, 126
   lime selection 108
   materials 37–94
   national standards 188–92
   simple methods 9
   soil stabilization 95–122
   summary 34
   three stage 34–5, 37–176
      stage 1 37–90
      stage 2 95–122
      stage 3 123–174
      stage 4 laboratory tests 36
   trial mixes 116
   see also lime reactivity tests
final mix, design 104–8
finishes
   floor 23–4, 162–5
   roof 173–4
   walls 22–4, 167–73
float 38, 156, 160, 164, 223
flocculation, clay particles 27, 98, 223
flood-resilient methods 1–2, 9, 14, *116*
flooding xiii–xv, xvi–vii, 8–11, 26
   mitigation 184
   prolonged 123
   regular 130–1

floors
 curing *166*
 earth floors 2, 5, 16–17
 finishes 22–4, 162–5
forts 4, 21–2
 Alhambra palace, Spain, x, 4, 15–16, 20–21, [photo 1c]
 Fort Jesus, Kenya, 22, [photo 1b]
 Kot Diji, Sindh, Pakistan 4, 20–1
 Fort Zanzibar 21, [photo 1a]
fossil 25, 224
foundations 123–132
 field-testing stage 3: 123–31
 of limecrete 123
 of lime-stabilized soils 126–7
 of stone 128–30
 with sandy soil 130–1
fuel burning in lime production 2, 6–7, 50, 54

Ghulam-Zaor, M. xvii
grading 224
 aggregates 189–90
 BS, BS EN, ASTM standards *190*
 soil 16, 98
 sand *see* particle sizes: sand analysis
granularity test 75
granulated blast-furnace slag (GBFS/GGBS) 91, 183
gravel 71–2, 74, 81, 127–8, 130, 189, 200
grout 65, 164, 224
Guillaud, H. xvii, 15
Gutschick, K.A. 179–80
gypsum (calcium sulphate) 15, 90–1, 224

handling
 care of block 132, 137
 cob 143
 quicklime 51 *see also* personal protective equipment; safety precautions
harling 150, 224
health and safety *see* personal protective equipment; safety precautions
health and sustainability 6–8

Hill, J. 30
historic buildings
 durability 3–5
 structures and materials 16–24
 traditional skills and current knowledge 13–15, 193
hoe (drag or hair hook) 37, *62*, 66, *67*, 224
Holmes, S. xi–xii 19
 and Rowan, B. 83
 and Wingate, M. xii, xvii, 10, 11, 172, 196, 199, 201
Houben, H. and Guillaud, H. xvii, 15, 27, 159, 162, 164
housing 19, 191, 193
 and community areas 185–6
hydration 103, 110, 225
hydraulic mix testing (summary) *175*
hydraulic set 26–30, 115–16, 225, [photo 11]
hydraulicity 29, 56

immersion test
 intervals 140
 Nepal [photo 17c]
 Pakistan *128*, 129, [photos 15a, 15b]
 pozzolans 84, 86, 92, 114
impermeability 6, 22, 162, 165
 see also vapour permeability; water: permeability
impervious 162, 225
impurities, active clay 199
Indian standards (IS) 189, 191
industrialized processes, plant and equipment (roads) 180–2
Innocent, C.F. 16, 18
Intergovernmental Panel on Climate Change (IPCC) 5
internal walls *see* finishes; plaster
ions 27

'jar test' *see* sedimentation test
joins, corners, junctions: render reinforcement 160

key (plaster, render) 65, 89, 141, 149, 152–3, 157, 167, 225–6
 see also lath; nibs
Kholucy, S. 89

kiln 2–3, 7, *48*, 226, [photos 8, 9]
  see also limestone: burning
knocking up 111, 132
kokoti (coral ragstone) 17, 19, 159
Kovan, E. 4

laboratory testing and analysis
  embankments, housing
      developments, roads,
      runways 119, 183, 186
  salts and sulphates 90–3
  soils and stabilization with lime 1,
      2, 5, 9, 16, 30, 32–3, 36, 86,
      104–5, 117, 196
  stabilization standards 189–92, 196
laminar structure 27, 226
laterites, Africa 177, 226
lath 149, 226
  see also key; walling techniques
      (earth): wattle and daub; nibs
leather hard 150, 157, 163–4, 227
lime 227
  acidity reduction 45, 93
  agricultural 58, 218
  air limes 218
    see also lime: non-hydraulic lime
  autogenous (self-healing) 179, 218
  bagged 47, 60, 219
  building limes 4, 7–8, 13, 37
    quality testing 47, 60–1
    see also national standards
  burns, treatments for 44–5
  cycle 25, *26*, 227, [photo 10]
    variations produced by hydraulic
        limes *31*, [photo 7]
  dead-burnt 54, 222
  dry hydrate (hydrated lime) 14,
      17, 224–5
    effects on soil 108, 109
    field tests
      and bulk density levels 60, *61*
      density testing 60
      fineness testing 60
      late hydration see late slaking
      late slaking 13–14, 147, 158
      pitting and popping 109,
          147, 230
    preparation and testing 59–61
    production 47, 50

  proportions in trial mixes 103
  sieve size *40*
  stabilization with 109, 113
  dust 46, 51–3
    see also safety precautions
  fat 223
  free 7, 224
  geological explanation see limestone
  hot-mixing 46, 53, 109–11, 113,
      172, 224
  hydrated lime see lime: dry
      hydrate
  hydraulic limes 28–30, 225
    artificial 29–30
    clay content of 199
    eminently hydraulic 29, 56, 59,
        61, 223
    feebly hydraulic 29, 59, 61, 223
    moderately hydraulic 29, 59,
        61, 228
    natural 28–9, 228
      dry hydrate from 60–1
      NHL2, NHL3.5, NHL5 229
      six-second test 56
      variations to the lime cycle *31*,
          35, [photo 7]
  lean 56, 58, 226
  lump 228
    see also lime: quicklime
  pit see lime putty: slaking tank and
      settlement pit construction
  pure 2, 7, 25, 55, 57–9
    see also limestone: pure
  putty see lime putty
  milk of lime 62–3, 71, 84–5, 228
  non-hydraulic lime 25, *26*, 229,
      [photo 11]
    forms of *47*, 108
    optimum lime proportions 97
    production of 47–50
    reactivity tests 55–9, 188–9
    and soil composition 73
    vs hydraulic lime 25, 27
    see also lime: cycle
  quicklime 14, 231
    air slaked 51, 60, 218
    powder 60
    crushing 38, *52*
    effects on soil 108–9

for external render 155
field testing 53–9
  observation test 53–4
  reactivity tests 57–9, 188–9
  six second test 55
  gain in weight measurement 56
hot-mixing 46, 109–11, 113
  for plaster 146–7
over-burnt 54–5
preparing for testing 51–3
production 47–50
proportions 102–3, 195–6
reactivity 46–7, 50–1, 53–4, 109–10, 128, 146, 188, 232
safety precautions and storage *49*, 51, 111
stabilization with 109–11
under-burnt 53–4
selection (form of) 47, 108–15
slaked lime 53, 56, 59–60, 85, 102, *110*, 147, 155, 161, 234
  see also lime: dry hydrate; lime putty
slurry
  civil engineering 181
  see also lime: milk of lime
strength gain see strength: long-term gain
lime concrete 72, 81, 130, *200*, 227
  see also limecrete; lime-stabilized soil concrete
lime powder see lime: dry hydrate; lime: hydraulic limes
lime putty 14, 47, 108, 227
  effects on soil 108–9
  for external render 155
  fattening see lime putty: maturing
  field testing 69–71
    consistency testing 70–1
    density testing 69–70
    fineness testing 71
    soundness testing 71
  maturing 68
  non-hydraulic 50
  production 66–8
  proportions in trial mixes 97, 103
    to pozzolan 141
    to soil 33, *102*, 103, 106, 111

safety precautions 42–4
slaking 234–5
  lime putty production *43*, 62–3, 66, *67–8*
  safety precautions 42–4, 67
  slaking tank and settlement pit construction 43, 61–5, *66*
  Nepal [photos 12a, 12b]
  stabilization with 109–10, 112
  storage 67
lime reactivity tests 57–9, 188–9
lime–sand mixes 14, 50, 73
lime-stabilized soil 235
  concrete 128
  see also limecrete
  pH 14, 58, 93, 103–4, 196, [photo 13]
  uses and benefits 1–12
limecrete 123, *131*, 227
  see also lime concrete
limestone 227
  Africa 177
  burning *2*, 6–8, 19, 25, 30, 47, *48*, 50, 54, 56, 60, [photos 8, 9]
  see also calcination
  geology 25, 29, 199
  pure 199
limewash 22, 23, 228
  aftercare 173–4
  application 169–70
  oil additives 89, 171–2
  as paint 167
  pigments 22, 171, 230
  preparation 167–9
linear shrinkage box test see clay content: field tests
linseed oil 22, 24, 89
Little, D.N. 178, 180
loam, definitions of 21
  see also clay: and loam
local sources and knowledge 73
long-term strength gain see strength: long-term gain
low-cost construction xiv–xv, 1, 8
low-cost field-testing equipment and materials xvi, 3, 7, 9–11, 30, 32, 40–1

machine block-making 135, 140, [photo 14]

marl 88, 228
materials
  choice of 32–3
  selection of 95–6
matrix 228
  hydraulic set 28
  trial render/plaster mix panels
    149–52, [photo 16a]
maturing
  lime putty 64, 68
  mixes 155
mechanized equipment 41, 197–8,
  [photo 14]
mellowing 98, 110, 186, 228
  industrialized process 180, 182
Memon, Q. xvii
Mendonça de Oliveira, M. 16
Mileto, C. and Vegas, F. 16
mineralogy 36, 61, 104, 160, 179,
  210, 228
mixes see trial mixes
mixing yard 37
moisture, protection of quicklime
  from 49, 51
Moisture Condition Value (MCV)
  test 183
moisture content testing 92, 133
mortars 19–20, 140–2, 228
  application 141
  bedding 13, 19–20, 65, 141, 152,
    162, 223
  damping and protecting 142
  magnified samples [photo 19]
  mix testing 141
  setting-time tests 141
moulds see block moulds

National Lime Association (NLA),
  USA 92, 177, 178, 180
  stabilization standard 192
national standards 11, 187–92
Nepal 1, 4
  compacted-earth block-making
    machine [photo 14]
  immersion testing [photo 17c]
  lime-slaking tanks [photo
    12a, 12b]
  plaster and render test panels
    [photo 16a]

render repair [photo 16c]
trial-mix samples [photo 17a]
Netherlands standard (N): pozzolans,
  191, 202–3
Neve, R. 21, 23
New Zealand Transport Agency
  (NZTA) 192
nibs 229, [photo 2]
Nigeria 177
nitrates 90
Norton, J. xvii, 10, 78, 115, 130, 134

oil
  and fats 15, 87–9, 172
    see also tallow
  floor finishes 164–5
  limewash finishes 89, 171–2
oiling box mould 134
organic additives 15, 87–9, 159–60

Pakistan
  compressive strength tests 119
  flood-relief programmes 1
  HANDS xi, xvii
  IOM xi, xvii
  immersion tests 128, 129
  immersion and compressive-
    strength tests
    [photos 15a, 15b]
  lime burning and production
    [photo 8]
  magnified mortar samples
    [photo 19]
  rice-husk ash 83
  selection of cured mixes from
    different locations
    [photo 17b]
  Sindh
    Kot Diji 4, 87
    Sukkur Barrage 30
pargeting 88–9, 167, 229
parging see pargeting
particle sizes 40, 72, 96, 229
  sand analysis 82–3
    see also sieves
Pearson, G. 18, 22
penetrometer testing 9, 83, 86, 119,
  128, 140, 205–6, 208, 229,
  [photo 15b]

permeability 229
　　see also vapour permeability; water permeability test
personal protective equipment (PPE) 41–6
　　see also safety precautions
pH see lime-stabilized soil: pH; soil: pH
pisé-de-terre 16, 22, 231
　　see also rammed earth
pits see lime putty: slaking tank and settlement pit construction; water storage and pit linings
pitting and popping 109, 147, 230
plaster 20–1, 146–9, 230
　　application and base coat 19, 71, 149, 151, 156
　　skim coat 13, 20, 234
　　testing and applying 149
plaster and render test panels 149–52, [photo 16a]
plasticity 88, 98, 108, 162, 201, 230
plasticity reduction: road building 179
plinths 128, 130–1, 143–4, 230
plumb bob 38, 141, 230
pointing 13, 20, 162, 230
Portuguese building techniques (16th-century Brazil) 16
potassium hydroxide 89
potassium sulphate 15
pozzolans 30–2, 83–6, 230–1
　　artificial 27, 83, 218
　　external render 159
　　field tests for reactivity 84–6
　　　　immersion and compressive strength 86
　　　　sedimentation 84–5
　　fly ash 83, 190–1, 223
　　indicative particle sizes 96
　　low-clay-content plaster 147–8
　　mixed designs 3, 10
　　national standards 190–1
　　as reactive minerals 83–4, 113, 147–8, 190
　　and soil composition 73
　　stabilization with lime and 113–14
　　strength gain: road building 179
　　testing 202–3
pozzuoli 32

precipitation 7, 231
Proctor test 183–4, 191
　　ASTM 192
protection (early stages) for lime work 37, 51, 65, 116, 153–6, 166, 172, 231
　　see also curing
protective clothing and equipment see personal protective equipment
pulverized fuel ash (PFA) 32, 121, 183, 185, 190, 229

quality testing and maintenance 47–51
quicklime see lime: quicklime

raking 66, 231
rammed earth 15–16, 144–5, 231
　　see also pisé-de-terre
rammers 135
　　see also CINVA ram
reabsorption (carbon dioxide) 7, 50, 115
reactive materials 95
reagent strip tests 92, 103–4, 232, [photo 13]
recording
　　mix ratios 121–2
　　and monitoring 117
　　test-mix record sheets 122, 207
Reid, K. 18, 146
render 21–2, 232
　　application and base coat 19, 71, 149, 151, 156–8
　　second and subsequent coats 157
　　skim coat 13, 20, 234
　　background viii, 22–3, 114, 150, 219
　　preparation 152–6, 160–1, 167
　　repair [photo 16c]
　　scratch-keying 152–3, 157, 167
　　trial mixes for additional durability 151, 158–62
　　wattle and daub 146
render and plaster test panels 149–52, [photo 16a]
repairs 1, 13, 17, 19, 187, [photos 3a–4b, 16c]

rice-husk ash 32, 83, *121*
road building *see* civil engineering
roller-pan mill 40, 198, 232
roughcast 150, 232
Rodney Melville and Partners xi, xvii
roof finishes 173–4
Rowan, B. xi
rural development 9, 11

safety precautions 41–6
   animals and children *44*, *53*, *61*, 62, 65
   clothing and equipment *42*, *45*
   dust 52–3
   external rendering 157
   hot mixing 53, 111
   quicklime *49*, 51, 111
   slaking 42–4, 67
Salmerón, P. 4, 15–16
salts and sulphates 90–3, 232–3, 235
   field testing 91–3
      conductivity 92
      observation 91
      reagent and indicator 92
      soluble salts 93
      swell 91–2
sand 72, 73, 233
   field testing 81–3
      particle-size analysis 82–3
   lime–sand mixes 14, 50, 73
   and loam 21
   selection 107
sandy soils 96
   foundations 126, 130–1
   plaster 147–8
scarification: industrialized process 181, 233
Schumacher, E.F. 193
scouring 24, 115, 150, 234
screeds 21–2, 65, 105, 115, *162*, 163–6
   roof 89, 173–4
   see also finishes; floors; water storage and pit linings
seashell 23, 25, 199
   see also calcium: carbonate
seismic 11, 119, 215–17, 234
   see also earthquake
shading *see* damping down; protection

Shaikh Tanveer Ahmed, Dr xvii
shrinkage 72, 78–9, 88, 91
   aggregate or fibre addition 72, 151, 159
   cob and rammed earth 144
   cow dung 88
   cracks in render 157
   disc test 78–9
   linear 207
   shrinkage box test *see* clay content: field tests
   quicklime use 109
   render 150
   swell 91, 200–1
shuttering 126, 128, 130, 144–5, 234
   settlement pit 63
sieves
   above settlement tank 63
   field tests for clay content 74, 77, 78, 79
   field tests for sand 81–3
   fineness testing 60, 71
   sizes *40*
silica 27, 32, 83, 234
size (glue) 22, 234
soak test *see* immersion test
Society for the Protection of Ancient Buildings (SPAB) 18
soil
   acidity 92–3
   composition 2, 3, 72
   field testing 71–9
   grading 98
   mortar *see* mortars
   pH 92–3, 103–4, [photo 13]
   see also reagent strip tests
   samples, obtaining 73–4
   sedimentation test 79–80
   selection 97
   stabilization *see* lime-stabilized soil
   suitability for addition of lime 73, 95–6, 189, 200–1
   tempering 98–9
step test 117–19
storage and protection
   external render mixes 155, *156*
   lime putty 67–8
   quicklime *49*, 51
straw 86–7
   see also fibres

Strawbuild xi, xvii
strength
 final 183–4
 long-term gain 8, 119, 144, 154, 179–80, 205–6
 time and temperature effects 180
 see also compressive strength
subgrade 10, 178, 184, 235
 strength 221
substructure 11, 178, 187
suction 231, 235
sulphates and salts 90–3, 235
Supakij, N. 184
superstructure 11, 178, 187
surface finishes *see* finishes
sustainability and health benefits 6–8
swell
 clay 72, *93*
 hydrate 55, 59, 103, 110
 with shrinkage *200–1*
 test 91–2

tallow 22, 89, 121, 159, 165, 172, 236
tamper 38, 115, *135*, 236
Taylor Wimpey: Gateshead residential development, UK 186
technical guidance
 civil engineering 182–4
 see also national standards
temperature
 autogenous healing 179
 burning 47, 54
 carbonation 28
 comfort 3
 curing 115–16, 138, 158, 169, 172–3, 180
 effect on strength 180
 fired brick 83, 113
 pozzolan test 202–3
 in quicklime reactivity measurement 57–9
 see also thermal conductivity; thermal insulation
tempering 19, 23, 98–9, 100, 128, 132, 147, 161, 236
tending 21, 115, 236
tensile strength 123, 236

test moulds 107–8
test-mix record sheet 122, 207
Texas Department of Transport 92
Texas Highway Department 119, 177
Thailand: Mun Bon dam 184
thatch roofing 174
thermal conductivity 3
thermal insulation 164
Thompsons of Prudhoe, England 186
Thorpe Hall, Suffolk 89
tiles
 floors 162
 pozzolan 83, 113
 roofing 89, 174
time
 boiling time measurement 57–9
 curing 50
 setting-time tests for mortar 141
 and temperature effects 180
toe *see* plinths
tools and equipment 37–41
TR Stabilization, England 186
traditional skills and current knowledge 13–15, 193
training programmes (flood resilience) xvi
transport and delivery: industrialized process 181
trass 86, 191, 203, 236
trench footings 72, 123–6, 185
trial holes and soil selection 74
trial mixes
 clay content estimation 100, *101–2*
 cob 144
 curing 115–16
 designing final mix 104–8
 field testing 116–21
 lime form selection 108–15
 preparation of materials for 97–9
 proportions for stabilization 102–4
 rammed earth 145
 recording and record format 121–2, 207–8
 render with additional durability 158–62
 samples 104–6, *107*, [photo 17a]

screeds and pit linings 165–6
selection of materials 95–6
tuff *see* trass
turpentine 23–4

UK
  Gateshead residential development 186
  historic buildings *see* England; Wales
  kiln [photo 9]
  road building 10, 177–8, 180, [photo 18]
    embankment 185
  see also Department of International Development (DFID); Department of Transport; *entries beginning* British
UN Climate Secretariat 5
unconfined compressive strength (UCS) *see* compressive strength
unslaked material 63, 71, 109, 111
un-stabilized mixes 114–15
USA
  civil engineering 177, 179–80, 184–5
  see also American Road Builders Association (ARBA); American Society for Testing and Materials (ASTM); National Lime Association (NLA)

vapour permeability 6–7, 21–2, 88, 159, 210–14
Vicat, L-J. 27, 202
vinegar/lemon juice for lime burns 45
voids 115, 200–1, 237
volcanic ash 32, 83, 237

Wales 17, 20–1
walling techniques (earth)
  adobe xv, 16, 18, 218
    trial mixes *see* blocks
  cob 142–4, 202
    floor finishes 23–4, 164
    trial mixes 144
  light-strawclay 211–13
  rammed earth xv, 15–16, *83*, *98*, *105*, 125, 144–5, 149, 189, 213, 231
    trial mixes 145
  straw bale iii, xi, 11, 211–13
  wattle and daub 18–19, 145–6, 237
    trial mixes 146
water 90
  content of trench footings and foundations 125
  dipping cured blocks before use 139
  'hose pipe' test 172
  permeability *see* water permeability test
  prolonged exposure 179
  purification 6, 26
  treatment of lime burns 44–5
  see also flood-resilient methods; flooding; immersion test
water permeability test 120–1, *Fig 5.18*
water shedding, prevention of 89, 171–2
water storage and pit linings 165
  protection and aftercare for screeds and pit linings 166–7
  trial mixes for screeds and pit linings 165–6
weathering qualities 237
Weismann, A. and Bryce, K. 23–4, 164
Welch, E. 19
William, E. 17, 21, 23
Wingate, M. 47, 57, 86, 202
  Holmes, S. and xii, xvii, 10–11, 172, 199
Wolfe Murray, M. xiii–xv, xvii, 7
wood burning *see* fuel burning in lime production
Woolley, T. et al. 6

Yates, J. 89

Zanzibar 13, 17, 19, [photos 1a, 5a, 5b, 16b]

www.ingramcontent.com/pod-product-compliance
Lightning Source LLC
Chambersburg PA
CBHW070913030426
42336CB00014BA/2396